JENNIFER LEIGH, JENNIFER HISCOCK,
ANNA MCCONNELL, CALLY HAYNES,
CLAUDIA CALTAGIRONE, MARION
KIEFFER, EMILY DRAPER, ANNA
SLATER, LARISSA VON KRBEK, KRISTIN
HUTCHINS, DAVITA WATKINS AND
NATHALIE BUSSCHAERT

WITH FOREWORDS
BY DAVID A. LEIGH AND SUE V. ROSSER

WOMEN IN SUPRAMOLECULAR CHEMISTRY

Collectively Crafting the Rhythms of Our Work and Lives in STEM

First published in Great Britain in 2022 by

Policy Press, an imprint of
Bristol University Press
University of Bristol
1–9 Old Park Hill
Bristol
BS2 8BB
UK
t: +44 (0)117 374 6645
e: bup-info@bristol.ac.uk

Details of international sales and distribution partners are available at
policy.bristoluniversitypress.co.uk

© Jennifer Leigh, Jennifer Hiscock, Anna McConnell, Cally Haynes, Claudia Caltagirone,
Marion Kieffer, Emily Draper, Anna Slater, Larissa von Krbek, Kristin Hutchins, Davita
Watkins, Nathalie Busschaert

The digital PDF and EPUB versions of this title are available Open Access and distributed
under the terms of the Creative Commons Attribution-NonCommercial-NoDerivatives 4.0
International licence (https://creativecommons.org/licenses/by-nc-nd/4.0/) which permits
reproduction and distribution for non-commercial use without further permission provided the
original work is attributed.

British Library Cataloguing in Publication Data
A catalogue record for this book is available from the British Library

ISBN 978-1-4473-6237-1 paperback
ISBN 978-1-4473-6238-8 ePub
ISBN 978-1-4473-6239-5 OA ePdf

The right of Jennifer Leigh, Jennifer Hiscock, Anna McConnell, Cally Haynes, Claudia
Caltagirone, Marion Kieffer, Emily Draper, Anna Slater, Larissa von Krbek, Kristin Hutchins,
Davita Watkins and Nathalie Busschaert to be identified as authors of this work has been
asserted by them in accordance with the Copyright, Designs and Patents Act 1988.

All rights reserved: no part of this publication may be reproduced, stored in a retrieval system,
or transmitted in any form or by any means, electronic, mechanical, photocopying, recording,
or otherwise without the prior permission of Bristol University Press.

Every reasonable effort has been made to obtain permission to reproduce copyrighted material.
If, however, anyone knows of an oversight, please contact the publisher.

The statements and opinions contained within this publication are solely those of the editors
and contributors and not of the University of Bristol or Bristol University Press. The
University of Bristol and Bristol University Press disclaim responsibility for any injury to
persons or property resulting from any material published in this publication.

Bristol University Press and Policy Press work to counter discrimination on grounds of gender,
race, disability, age and sexuality.

Cover design: Bristol University Press
Front cover image: Jennifer Leigh

Printed and bound by CPI Group (UK) Ltd, Croydon, CR0 4YY

This book is dedicated to Sue Hiscock

Contents

List of figures and vignettes		vi
Notes on contributors		viii
Acknowledgements		xiii
Foreword by David A. Leigh		xiv
Foreword by Sue V. Rosser		xvi
one	Introduction	1
two	A qualitative approach: autoethnography and embodiment	16
three	Building academic identity in the context of STEM	33
four	Challenges for women and marginalised groups	52
five	WISC: Women in Supramolecular Chemistry	69
six	Stories from STEM	90
seven	For the future	109
References		124
Index		153

List of figures and vignettes

Figures
1.1	What is a chemist?	10
2.1	Which path should I take?	30
3.1	Being an early career researcher is like fighting a giant squid	36
3.2	The constant question of how to time family into the career motorway before the stork crashes into the brick wall of age	41
3.3	Carrying it all	46
3.4	My ideas being 'trimmed' to only keep the few viable and profitable ones	49
4.1	Crushed by chemistry	61
4.2	COVID-19 'Can you mind the baby?' 'Buried and burned'	66
5.1	Interweaving threads of WISC projects	79
5.2	Design for enamel badge incorporating the WISC logo	80
6.1	'Unprecedented' sudden emergency	93
6.2	Being online	94
6.3	I'm a small dot between two giant rocks	96
6.4	Why is it so hard to put myself first?	99
6.5	The benefits of working at home	100
6.6	This is how I feel sometimes, one flower having to stand tall while everything else is dying around me	102

Vignettes

First vignette	Maria, 32, Early career	31
Second vignette	Adi, 23, International PhD student	50
Third vignette	Paula, 33, Early career	67
Fourth vignette	Mira, 27, Post-doc	88
Fifth vignette	Hermione, 38, Mid-career researcher	107
Sixth vignette	Phyllis, 63, Senior researcher	122

Notes on contributors

Nathalie Busschaert is Assistant Professor at Tulane University (New Orleans, USA). She obtained her BSc and MSc in chemistry at the Katholieke Universiteit Leuven (Belgium) and a PhD in supramolecular chemistry from the University of Southampton (UK) under the supervision of Professor Philip A. Gale. After her PhD, Nathalie continued research in supramolecular chemistry under the supervision of Andrew D. Hamilton at the University of Oxford (UK) and New York University (USA). She started her independent career at Tulane University (USA) in 2017. The Busschaert group is working on molecules that can selectively bind to lipid headgroups and is exploring various applications of transmembrane transport.

Claudia Caltagirone is Associate Professor of Inorganic Chemistry at the University of Cagliari (Italy), working on the supramolecular chemistry of anion and metal ion recognition and sensing and on self-assembled supramolecular architectures. She obtained her PhD in chemistry in 2006 at the University of Cagliari under the supervision of Professor Vito Lippolis and then she moved to the University of Southampton for two years as an academic visitor in the group of Philip Gale. She is a co-founder of WISC and in September 2021 she was in charge of the first WISC Workshop at the University of Cagliari.

NOTES ON CONTRIBUTORS

Emily Draper received her PhD from the University of Liverpool in 2015 with Professor Dave Adams. She carried out postdoctoral research at Liverpool and then at the University of Glasgow. She became a Leverhulme Trust Early Career Fellow and a Lord Kelvin Adam Smith Fellow in 2017. She took up a lectureship position at Glasgow in 2018, working on self-assembled organic electronics, before taking maternity leave in 2019 and 2021. She is a UKRI Future Leaders Fellow.

Cally Haynes is Lecturer in Organic Chemistry and Chemical Biology at UCL, UK. She studied chemistry at the University of Oxford and completed a PhD and initial postdoctoral work in supramolecular chemistry at the University of Southampton under Professor Phil Gale. In 2013 she left academia and began working in scientific publishing with the Royal Society of Chemistry. However, after missing the lab and supramolecular research she came back to the field in 2015, carrying out postdoctoral work at the University of Cambridge until 2019, when she moved to London to establish her own group for the first time. She is a co-founder and Vice-Chair of WISC, currently coordinating the website and online resources.

Jennifer Hiscock is Reader in Supramolecular Chemistry, Chair of the international Women in Supramolecular Chemistry (WISC) network, and UKRI Future Leaders Research Fellow within the School of Chemistry and Forensic Science at the University of Kent (UK). She studied for a PhD in the group of Professor Phil Gale at the University of Southampton (now Sydney) and carried out postdoctoral research in the same group until 2015. Following this, she moved to the University of Kent as the Caldin Research Fellow and was appointed Lecturer in Chemistry at this same institution in 2016. Her current research interests focus on the development of Supramolecular Self-associating Amphiphiles (SSAs) as antimicrobial, anticancer, and drug adjuvant agents.

Kristin Hutchins is Assistant Professor of Chemistry at Texas Tech University (USA). Kristin studied chemistry at the University of Iowa, where she earned her BS and PhD. Kristin obtained her PhD in 2015 while working under the supervision of Leonard MacGillivray. She conducted postdoctoral research at the University of Illinois at Urbana-Champaign with Jeffrey Moore. Kristin began her independent career in 2017, and the Hutchins group focuses on using supramolecular chemistry and crystal engineering strategies to control the properties of solid-state organic materials.

Marion Kieffer is R&D Scientist at InnoMedica, a small Swiss pharma company, working on liposomal drug formulations. She made the tough call to leave academia behind after completing a PhD and post-doc in supramolecular chemistry in the UK.

Jennifer Leigh is Senior Lecturer at the Centre for the Study of Higher Education, University of Kent. She initially trained as a chemist, somatic movement therapist, and yoga teacher before completing her doctorate in education at the University of Birmingham. She edited a book for Routledge in 2019, *Conversations on Embodiment across Higher Education: Teaching, Practice and Research*. Together with Nicole Brown, she edited and contributed to *Ableism in Academia: Theorising Disabilities and Chronic Illnesses in Higher Education* (UCL Press, 2020), and authored *Embodied Inquiry: Research Methods* (Bloomsbury, 2021). She is a founder member of WISC (an international network for Women in Supramolecular Chemistry) and as Vice Chair (Research) is the only social scientist on the Board. The projects she has underway with WISC aim to bring embodied research approaches (and glitter) into the world of chemistry and public engagement. Her next book, *The Boundaries of Qualitative Research: With Art, Education, Therapy and Science*, will be published by Bristol University Press. Her research interests include embodiment, phenomenological and creative research methods, academic practice, academic development, and ableism as well as aspects of teaching and learning in higher education. Her email is j.s.leigh@kent.ac.uk. She tweets as @drschniff @suprachem.

Anna McConnell studied chemistry at the University of Canterbury in New Zealand before obtaining a DPhil under the supervision of Professor Paul Beer at the University of Oxford. Following postdoctoral research stays at the California Institute of Technology and the University of Cambridge in the groups of Professor Jacqueline Barton and Professor Jonathan Nitschke, respectively, she became a Junior Professor at Christian-Albrechts-Universität zu Kiel in Germany in November 2016. The McConnell group's research focuses on stimuli-responsive metal-organic cages, dynamic covalent chemistry, and luminescent complexes. She is a co-founder of WISC and Vice-Chair (Organisation).

Anna Slater is Royal Society University Research Fellow at the University of Liverpool. She obtained her PhD from the University of Nottingham in 2011 in the group of Professor Neil Champness. Following postdoctoral positions in porphyrin self-assembly and organic materials, she took up a Royal Society-EPSRC Dorothy Hodgkin Fellowship in 2016. She started her current fellowship in January 2021. Her research interests include using molecular design and process control for the formation of bespoke supramolecular architectures, organic materials, and macromolecules. She joined WISC as a Vice-Chair in 2020. She has two daughters (three and six), has a hidden disability, and tweets as @AnnaGSlater.

Larissa von Krbek is Emmy Noether Junior Research Group Leader at the Kekulé-Institute for Organic Chemistry and Biochemistry, University of Bonn, Germany. She studied chemistry at Freie Universität Berlin, Germany, and obtained her doctorate under supervision of Professor Dr Christoph

A. Schalley at the same institution in 2016. Following a postdoctoral stay in the group of Professor Jonathan Nitschke at the University of Cambridge, she moved to the University of Bonn to start her independent career in 2020. The Krbek group is working in the areas of supramolecular chemistry, systems chemistry, and out-of-equilibrium self-assembly. Larissa is a Vice-Chair of WISC working on communications together with Cally.

Davita Watkins is Associate Professor of Chemistry at the University of Mississippi (USA), where her research interest is in developing supramolecular synthesis methods to make new organic semiconducting materials for applications in optoelectronic devices, as well as studying their structural, optical, and electronic properties. Her group also investigates the design of dendrimer molecules for biomedical applications. Her research allows her group to use tools from all areas of science including analytical, computational, and materials chemistry. In addition, they get to enjoy collaborating with a number of theoreticians and other experimentalists.

Acknowledgements

WISC would like to thank all its funders and supporters (University of Kent; University College London; Università degli Studi di Cagliari; Christian-Albrechts-Universität zu Kiel; Royal Society of Chemistry: Inclusion and Diversity Fund; UKRI Future Leaders Fellowship, Grant Number: MR/T020415/1; Royal Society, Grant Numbers: APX\R1\201170; APX\PE\201004; Scot Chem; Biochemical Society: Diversity in Science Grant; ChemPlusChem; Crystal Growth & Design; STREM Chemicals), the Board and Advisory Board members, Sarah Koops who worked as a Research Assistant, and all the wider members who have participated in the clusters, the mentoring programme, attended events, and engaged with us on Twitter. We could not have done this without you.

WISC would also like to thank all those who have participated in our research, and promise that no alpacas were harmed in the process.

In addition, JL would like to thank everyone in WISC for trusting her enough to come on this journey with her; her family, Jimmy, Kira, Summer and Lincoln, for (still) allowing her to appropriate the sofa as her office and bringing endless encouragement, support, and hot and cold drinks; her parents, Anne and Jon, and her late father-in-law Jeffrey for believing in her; and her friend Elliott for being her workmate through the bulk of the writing and being her cheerleader.

Foreword

David A. Leigh FRS

Royal Society Research Professor and Sir Samuel Hall Chair of Chemistry, The University of Manchester

"What is WISC?", I asked a friend as we shuffled into the lecture theatre for a lunchtime presentation at a conference a few years back. At that presentation I learned that WISC is an organisation of (mostly) women (mostly) academics whose aim is to promote awareness and inspire action to address issues of equality and diversity in the supramolecular chemistry community. This is their first book. And for me, a white male professor in my late 50s, a Fellow of the Royal Society, it is an uncomfortable, unsettling, but important read. If I have to work late or leave when it's dark, I do so without a second thought. If I'm invited to a conference that I wish to go to, I can generally go. If I have a grant or paper rejected, or find out that I earn less than the professor next door, I never have any doubts as to whether my gender had anything to do with it. That's simply not the case for women in my field, or in my profession generally. It's difficult – perhaps impossible – for me to fully appreciate what that's like, how it impacts one's career and confidence, how it affects day-to-day life. But this book brings it closer; one feels the authors' frustration and shares their rightful sense of injustice. It is clear from the book that WISC also works as an effective support network for its members. It was great to learn that everyone who has taken an active role in the organisation has seen an increase in research outputs, grant successes, or career progression.

Over my career I have seen many things change for the better in academia: Recruitment and promotion committees take genuine steps to avoid conscious and unconscious bias; schemes have been introduced that target women and other disadvantaged groups for independent positions; the increase in the number of women in chemistry departments has drastically changed the 'macho' culture that was prevalent 25 years ago. But the text and vignettes in this book, the latter composed from real experiences of women in supramolecular chemistry, paint a vivid, troubling picture that shows just why further significant change is still needed. The playing field is still not level. Whether that's the fault of society, academia, or supramolecular chemistry itself, I don't know. But I suspect it's all three.

In reading this book, the most uncomfortable part of all was the persistent wondering if and how my own behaviour contributes to the inequality and experiences I was reading about. What do I do, or not do, that makes academia less fair on my women colleagues? And my questioning of that is, perhaps, the best reason of all for this book.

What is WISC? It's a start.

Foreword

Sue V. Rosser

Provost Emerita at San Francisco State University and Professor Emerita of the School of Public Policy at Georgia Institute of Technology

Women in Supramolecular Chemistry documents the exciting journey of discovery and recognition of a community of women scientists in this subfield of chemistry that remains male-dominated even today. The community originated from a small group of friends and peers who met bi-weekly to mentor and support each other in their research, grants, and publications. It expanded to an established network to examine diversity, equity, and inclusion issues in supramolecular chemistry and support retention and progression of post-PhD women in the field with data, numerous projects, and financial support through the Royal Society of Chemistry Diversity and Inclusion fund, as well as from several grants. Its emergence, over the relatively short period of 2–3 years that included a global pandemic, makes this an especially heady, compelling, and timely story.

In many ways, the development and history of Women in Supramolecular Chemistry (WISC) shares the phases of progression and growth experienced by other movements such as those of Civil Rights, women's, LGBTQIA, disabled, and other communities for recognition and equality. Scholars have evolved a variety of phase models to explicate the development of the movements in general, as well as their particular evolution within the academy. Building upon the phase model of McIntosh (1983), Schuster and Van Dyne (1985) and Tetreault (1985) for transformation as the result of women's studies, I developed (Rosser 1990) such a model tailored to the STEM fields. This model included the following six stages:

Stage 1: Absence of women not noted. This is the traditional approach to science and the curriculum from the perspective of the white, Eurocentric, middle-to-upper class male in which the absence of women is not noted and gender affects neither who becomes a scientist nor the science produced.

Stage 2: Recognition that most scientists are male and that science reflects a masculine perspective on the physical natural world. A few exceptional women such as Nobel laureates who have achieved the highest success as defined by the traditional standards of the discipline may be accepted in the scientific community and included in the curriculum.

Stage 3: Identification of barriers that prevent women from entering science. Women are recognised as a problem, anomaly, or absence from science and the curriculum. Women may be seen as victims, as protesters, or as deprived or defective variants, who deviate from the white, middle-to-upper class norm of the male scientist.

Stage 4: Search for women scientists and their unique contributions. The extent to which the role of women has been overlooked, misunderstood, or attributed to male colleagues throughout the history of science is explored to determine women's scientific achievements.

Stage 5: Science done by feminists/women. In this phase, new perspectives result when women become the focus. Topics chosen for study, methods used, and language in which data and theories are described may shift and become expanded, improving the quality of science.

Stage 6: Science redefined and reconstructed to include us all.

In several aspects, the development of WISC over the last three years, and charted in this volume, follows this phase model. Not surprisingly, Stage 1, in which the absence of women was not noted, does not apply in the same way to WISC in the 21st century as it did to the consciousness-raising groups of the late 1960s and 1970s that gave rise to the movements in ethnic studies, women's studies, disability, and gay/lesbian/queer studies. However, as described in Chapter One and especially in Chapter Five, WISC did begin from a group of young early-to-mid career women friends and colleagues in supramolecular chemistry who met bi-weekly to share information and mentor each other in their careers because they felt relatively isolated due to the absence of women in the subdiscipline.

Despite seeking advice and receiving some support from senior women in the field, they remained clear that the initial group would focus on issues for early- and mid-career women. Similar to the women and underrepresented people of colour in STEM in the US in the 1970s who realised that they needed data to understand their underrepresentation and the reasons for it, and sought a congressional mandate for the National Science Foundation (2000) to collect data and publish biennial reports beginning in 1982 with the *Status of Women and Minorities,* to which *Persons with Disabilities* was added in 1984, the group of colleagues realised in 2019 that they needed to collect data on the situation of women in supramolecular chemistry. They added a social scientist to their group to help them explore new methods and develop a survey to reach out to others in the field, as they describe in Chapter Five.

The co-authors begin the introductory chapter by emphasising that the volume does not centre on barriers to careers in STEM, or women in supramolecular chemistry as victims, or on blaming men (Stage 3). In Chapter Five, when they describe the history of WISC, they underline this again in their emphasis upon 'calling in' rather than 'calling out'. However, parts of Chapter One and much of Chapter

Four focus on issues or problems women in supramolecular chemistry face in establishing and maintaining momentum in their careers; the vignette at the end of Chapter Three describes a woman who decides to leave the field. In some ways, Chapter Two, the methodology chapter, in which the authors describe their reasons for choosing methods of autoethnography and embodiment, because they recognise the limitations of the traditional methods of supramolecular chemistry, represents a recognition of a problem of the limitations of the field, particularly for women who seek a diversity of methods to understand the world.

The authors have sought the data on the numbers and accomplishments of women in supramolecular chemistry (Stage 4) and provide some names and references to successful senior women in the subdiscipline, especially in Chapter Five. However, just as they do not set out to focus on barriers, they also limit this information, since it was not their intention to provide a history of women in supramolecular chemistry.

Quite appropriately, the authors devote much of the volume, especially Chapters Three and Six, to the challenges and opportunities these early- and mid-career women currently face as they build and sustain their careers, particularly in the face of a global pandemic. Both in their feminist approaches and in their search for new methods such as autoethnography and embodiment, use of drawings, vignettes, and other creative arts-based methods, as well as in their language of inclusion, the authors have written a Stage 5 book. In my opinion, this is what makes the volume especially interesting and exciting.

Their ultimate goal, which they state in the introductory chapter, and emphasise heavily in Chapter Five, as well as intermittently throughout, is inclusion of all (Stage 6). They are clear about wanting to include men, more senior women, and others in the three community clusters they have established around parenting, disability/chronic illness/neurodivergence and first-generation supramolecular chemists. In several instances they reiterate their commitment to 'calling in' rather than 'calling out', in recognition of making a better, more inclusive science for all.

The authors interweave emphases on diversity and intersectionality among women, the significance of other movements such as #MeToo and neurodivergence throughout their telling of the lived experiences of these women supramolecular chemists. These emphases mark another distinction between this and the mid-20th-century movements. The international inclusion and reach of the group, beginning in Europe, but now spreading to Asia, Australia, North America, and Africa, particularly distinguishes these women in STEM, as does its use of creative arts-based, autoethnography and embodiment methods. The creative methodologies of the drawings and vignettes interspersed throughout reveal and point out the similarities and differences arising from different international locations.

Other than its focus on women in supramolecular chemistry, perhaps the most significant difference between this volume and others based upon the lives and careers of women in STEM is that it chronicles experiences of women scientists during the global COVID-19 pandemic. The women's revelations of struggles to parent while working remotely, maintain a functioning team to conduct research when some or none could be in the laboratory, and sustain long-distance relationships with partners and families when travel, particularly across international borders, was not possible, provide new insights into how research, careers, and families have been impacted by the pandemic. These insights and information make *Women in Supramolecular Chemistry* especially important reading for all in supramolecular chemistry and science, especially male scientists, university and research administrators, funders, and reviewers, as well as other women scientists.

References

McIntosh, Peggy (1983) *Interactive Phases of Curricular Development: A Feminist Perspective.* Wellesley, MA: Wellesley College Center for Research on Women.

National Science Foundation (2000) *Women, Minorities and Persons with Disabilities in Science and Engineering* (NSF 00–327). Arlington, VA: National Science Foundation.

Rosser, Sue V. (1990) *Female-Friendly Science: Applying Women's Studies Methods and Theories to Attract Students.* New York: Pergamon Press.

Schuster, Marilyn R. and Van Dyne, Susan R. (1985) *Women's Place in the Academy: Transforming the Liberal Arts Curriculum.* Totowa, NJ: Rowman & Allenheld.

Tetreault, Mary K. (1985) Stages of thinking about women: An experience-derived evaluation model. *Journal of Higher Education*, 5(4), 368–384.

ONE

Introduction

Introductions commonly set the scene for a book, and let the reader know what they can look forward to, what will be included in the content, what the book is and who it is aimed at. Before starting that more conventional introduction, however, first we want to set out what this book is not. This book is not an account of how hard women and other marginalised groups have it in science. It does not contain victim stories or whistleblowing from women in supramolecular chemistry who have an axe to grind and want to call out all the men who have been mean to them. **We should note at this point that in this chapter we do discuss sexual harassment and this may be distressing for those who have experienced it**. We recognise that whistleblowing is a courageous act done often by those who have been subjected to trauma, and we respect, support, and thank all those who share their often difficult and traumatic stories. However, we wanted to go about things differently. A book of victim stories might gain traction, it might be 'clickbait' for those who want to read a tell-all about life in science, but it would do nothing for the careers of those who wrote it and those it seeks to help. Confessional books such as that, stories such as those, tend to be shared by women towards the end of their careers, when they have left science, or when they have retired – see, for example, work by Rita Colwell[1], Ellen Daniell[2], Sue Rosser[3–5], and Vivian Gornick[6]. These stories are important and very much need to be told. However, women at that point in their life have little to lose by calling out sexual harassment, recounting tales of being ignored, overlooked, and disregarded. In contrast, the authors here are all women in their early-to-mid career stages, who are working (with the one exception of our social scientist) to establish themselves as researchers and leaders in their field of supramolecular chemistry. They are women who are committed to changing things, to making things better for themselves and those women who will follow them, and whose time is spent doing and researching science in laboratories and in their discipline. We 'call in' the community to support those who are marginalised,[7] and share how this might bring about positive change. This is not to say we are whitewashing the very real challenges faced by women and other marginalised genders.* Drawing on the example of Black feminist researchers: 'To only focus on the strengths, accomplishments and victories does not give sufficient attention to the system of domination'.[8(pxii)] We want to walk in the footsteps of the women who have walked before us,[9] be agents of change, and continue to bring awareness to the idea of gender and other marginalisations within science.

Who this book is for and why it matters

This book is for everyone who works in supramolecular chemistry, chemistry, and science. It is for the university and research administrators, the funders, the reviewers, and all the people who make the decisions that are responsible for

---

* Our use of 'women' includes trans women. We use 'marginalised genders' to also include non-binary people and trans men.

marginalising women and others. As the majority of these people are still men, this book is for those men, so that they can change the balance of gender in science.

Why do we need to talk about gender in relation to science? Why do we *still* need to talk about gender in relation to science? Isn't this old hat now? Sara Ahmed wrote, 'it can be deemed more old-fashioned to point out that only white men are speaking at an event than to have only white men speaking at an event'.[10(p155)] Similarly, is it more old-fashioned to point out that there is a need to talk about the lack of gender balance and diversity than it is to just carry on as we are? To an extent this is true. It *can* seem old-fashioned and regressive to keep banging on about gender. But there is a need for change. In 2019, the Royal Society of Chemistry wrote that at the current rate of change we would never reach gender parity in the chemical sciences.[11] There has been progress since the 1960s when 'chemistry department heads said as openly as they had in 1940: "we don't hire women"'.[6(p88)] By the 1980s in the same department they said: 'The chemistry department here doesn't advertise. It's illegal now, but they still do it that way. Somehow, they consider it a "shame" to advertise. They write to their friends. And of course their friends are men who have only male graduate students'.[6(p88)] Things have definitely moved on from then; numbers of women undergraduates, postgraduates, and faculty have increased since the 1970s,[5] although 'the cutting edge of science and engineering remain out of the reach of the vast majority if not all women'.[5(p49)] In 2008, Aviva Brecher looked back over her career and expressed shock about the women starting their scientific careers: 'it is astonishing that they are asking today the same questions about how to successfully manage and blend careers in science with the demands of motherhood and family life we struggled to solve thirty years ago'.[12(p25)] However, choices and challenges balancing life are not the only ones that a woman working in STEM (science, technology, engineering, and mathematics) will encounter.

The struggles faced by women and other marginalised genders are not new, and many of these struggles are based around how their bodies are seen and treated in the world. In 2006, Tarana Burke initiated the #MeToo movement on social media. Its original intention was to empower women with a sense of solidarity and through strength in numbers. In 2017, this hashtag was adopted by Alyssa Milano in response to the allegations of sexual misconduct by Harvey Weinstein. She wrote on Twitter: 'If all the women who have been sexually harassed or assaulted wrote "Me too" as a status, we might give people a sense of the magnitude of the problem'.[13] Women began using it to talk about experiences of sexual assault, including high-profile women across a multitude of sectors. The hashtag trended in over 85 countries. In 2021, the website Everyone's Invited[14] caught the attention of the media in the UK.[15–17] A place where women were asked to share testimonials of abuse they had experienced within institutions of education and/or work, dedicated to eradicating rape culture through conversation, the site had attracted 14,000 testimonies by the end of March 2021, and 51,000 by the middle of July 2021.

Women are not generally encouraged to speak out about their experiences. According to Rape Crisis 20% of women have been assaulted since the age of 16,[18] but conviction rates are shockingly low when compared to other crimes.[19] Dame Vera Baird, Victims' Commissioner for England and Wales, said: 'We have seen a seismic collapse in rape charging and prosecutions', with only 1.6% of reported rapes resulting in a charge or summons for the perpetrator.[20] However, as might be gathered from the response to #MeToo and www.EveryonesInvited.uk, it is likely that even these official statistics underplay the situation. According to the Brennan Centre for Justice, in the US up to 80%[21] of rapes may go unreported, and in the UK up to 85%.[20] Sexual assault and sexual harassment cover a wide gambit of crimes, including rape and much, much more. In a context where women are routinely blamed for 'asking for trouble' or putting themselves into dangerous situations such as walking outside in the dark, wearing clothes that are too revealing, or drinking too much, lived experiences of harassment and abuse are endemic, and this as true for women who work in STEM as in any other field.

Women are used to being blamed, being questioned, having their legitimacy, their choices, and their authority interrogated in many aspects of their day-to-day lives. Many have internalised misogyny, in the same way that many disabled people internalise ableism[22] and people of colour internalise racism.[23] Since classical times, there has been a tacit acceptance that public spaces belong to men.[24] The very suggestion, made by Jenny Jones in the House of Lords, that men should have a curfew as they are responsible for 97% of all violent attacks resulted in 'a massive misogynistic hissy fit of outrage. Most were completely blind to the ways that women have to constantly adapt their lives in reaction to male violence'.[25] Women, on the other hand, belong in the private spaces, the domesticated spaces. And there they should stay. They are wives, mothers, sisters, daughters. Their worth is determined by their relationship to the men around them. Women and feminism have done much to challenge this culture, but at the time of writing the patriarchy and male violence towards women remain a pervading and pervasive truth which impacts how women live their lives.

Women in STEM, women in science, and women in chemistry are still women. When we talk about numbers of women in scientific disciplines, when we talk about neoliberal cultures of overwork and hyperproductivity, when we talk about isolation and the loneliness of women working within environments that are dominated by men, we cannot ignore that these women are still women, whether they have a womb or not. When a woman is choosing whether to work late in the lab, she has to do so with the knowledge that it is not *just* about a choice to work late. It is a choice to work late in a room or building where there is minimal presence of others (and those who are there are likely to be male). It is about the choice to walk from that building alone, and to make her way home alone, walking, on public transport, or in a car, and then to enter her home, alone. These choices will be mediated by her identity, and whether in addition to being a woman she is trans, neurodivergent, a woman of colour, or has a disability, for example. Each of these intersecting identities present additional risk, additional factors that may balance her choice to work late, or leave while it is still light, or when others are leaving. This train of thought may seem trivial to any man who only has to weigh up the choice of working late or not despite the vast majority of violent attacks being male-on-male violence. However, as events such as those of the week of 15 March 2021 in the UK highlighted,[26] together with ongoing campaigns such as #ReclaimThe Street, for a woman they can be a matter of life and death, or at least a life free from the threat of rape or sexual assault (this particular week included International Women's Day; a highly publicised interview with a former princess revealing mental health issues; the discovery of the body of a white woman murdered in London by a serving Metropolitan Police officer; victim-blaming in the media; disruption of peaceful vigils by the Metropolitan Police; and Mother's Day). When we look at the history of women in science, and we look at the perspectives on why there is a lack of gender balance, as well as discounting outmoded beliefs that women just are not capable of cutting-edge research,[6] we also need to address the context within which women in science operate. If we are to look comprehensively at the experiences of women in science, then we need to foreground their experiences as women.

Women in science

The numbers show that there has been bias and marginalisation of many different groups in science; this is indisputable.[5,11,27,28] This is not due to lack of scientific interest by women: 'women have always been interested in science. The fact is, women have been actively excluded from science'.[1(p182)] Since Victorian times, attitudes and beliefs that women were inferior to men made it more difficult to pursue a career in science.[29] Once thought of as a profession for the 'sons of educated men', now it is widely recognised that science needs to be diversified because:

> the best of 100 percent of the population will always be better than the best of 50 percent of the population. Once all the talent in our country can compete on a level playing field, decisions about who to hire and who to support can be made on the basis of brains and ability, not gender, ethnicity, or national origin.[1(p194)]

In order to get the very best and brightest working on the huge problems that face our world today – global warming, global pandemics, hunger, poverty, and health to name a few – then we need to be fishing in a pool that includes everyone. This book goes beyond these numbers that demonstrate so clearly that there is a problem in science. Those numbers have been recorded and reported at least since the 1970s, and yet the pace of change has slowed and stalled:

> The percentage of women on the faculty of MIT's [Massachusetts Institute of Technology] School of Science went from zero in 1963 to 8 percent in 1995 to 19.2 percent in 2014. But since then, the drive toward equality has stalled. There were fourteen women faculty members in MIT's biology department in 2009 – and there were still fourteen ten years later in 2019. The proportion of women faculty in biology and chemistry actually decreased during that time.[1(p75)]

Gender equality has been identified as a goal worthy of global challenge by the United Nations.[30] The need to address equality diversity and inclusion in science has been foregrounded by the NSA (National Science Academy), UKRI (UK Research & Innovation), and professional bodies, even before the 2020 pandemic and its impact on gender balance.[31,32]

This book seeks to humanise the experiences of those women in STEM, reaching out with embodied stories to evoke responses as to *why* this lack of gender balance and diversity matters, why it is important, and what impact it has on those who are affected by it. We are all, as authors, passionate about this topic. Throughout, we keep 'a strong presence of emotional investment visible'.[33(p23)] In order to do this, we include images created during and for our research along with *vignettes* throughout each chapter. The *vignettes* are fictionalised composite stories that have been written in response to the findings of WISC's ongoing research (see Chapter Five for the story of WISC and its current projects and Chapter Two for an exploration of the methods and approach used). They are used to highlight the personal stories and lived experiences of women in supramolecular chemistry in a way that does not allow individuals to be scapegoated, blamed, or identified in any way. This is an effort to avoid the repercussions on individuals as a result of 'whistleblowing' or complaining.[34] None of the stories told in the *vignettes* are 'real', but all are true, and draw on lived experiences.

The stories of women in STEM have been a subject of interest for a long time, from women scientists in the First World War,[29] to reminiscences from women looking back at their careers in science in the 1980s.[6] This book adds to that literature, and differs from it in two important ways. First, it takes a field-specific approach. While the issues and experiences that we highlight here are common to many women in STEM, STEM is not a monolith: it comprises/represents many different disciplines. Any conclusions drawn from data collected will depend upon the agency collating it and which disciplines are included in that agency's investigation. Some disciplines and fields have made much more progress towards equality, diversity, and inclusion than others in the last 50 years, and therefore the discrimination faced by women and other marginalised groups is not the same in each discipline. For example, some studies include the physical sciences, others separate out psychology, and yet others include psychology and medicine, which both have a good representation of women. The discrimination faced by women and other marginalised groups is not the same in each discipline. This book focuses in on one field of chemistry – supramolecular chemistry – and the perceptions and experiences of those who work within it (see Figure 1.1, which shows one author's response to the prompt 'who are you as a chemist?').

Second, this book is not written *about* women in STEM. It is written *by* and *with* them. The authors all took part in work that set out to use creative and embodied approaches to capture stories, build community, and identify a toolkit for women who want to progress in the field. The timing of this project coincided with the 2020/21 COVID-19 pandemic and lockdown, and as such records the experiences of women scientists as they navigated through life inside and outside laboratories. The authors range in age from their 20s to mid-40s, and represent the range of early, mid, and senior career scientists with and without children. The topics that they discussed and explored led to themes that are included in this book: the reality of being a woman in a male-dominated sphere; the fight for lab space; the pressure and responsibility towards their teams. They represent a diversity of race, ethnicity, religion, nationality, and disability. They include those who chose to leave the chemical sciences and academia in addition to those who chose to remain. This work was supported through WISC, an international network for Women in Supramolecular Chemistry, with the aim to support equality and diversity.

WISC aspires to be an agent of change. It launched in November 2019. Since its launch the network has created a website and resource bank[35] (www.womeninsuprachem.com),

Figure 1.1: What is a chemist?

All of the images in this book are the authors' own.

conducted a survey of the supramolecular community, initiated small-group mentoring, and published a paper outlining the ethos of 'calling in' the community and the network's approach to combining qualitative research methods and expertise in EDI (equality/equity, diversity, and inclusion), together with findings from the first survey of the supramolecular chemistry community in *Angewandte Chemie*, a leading international chemistry journal,[7] and from a second survey on lived experiences through COVID-19 in *CHEM*.[36] The network has set up an international board and advisory board. It has led a series of panel events (in-person pre-COVID and online) in collaboration with MASC (the Royal Society of Chemistry special interest group for Macrocyclic and Supramolecular Chemistry) and virtual MASC (the online, early-career arm of MASC). WISC's first skills workshop for early-career researchers was originally organised in September 2020, but was postponed to September 2021 in Cagliari.[37] The WISC has set up support clusters – groups where individuals can share experiences and support and learn more about supporting others. These include a parenting cluster, a disability, chronic illness, and neurodivergence cluster, and one for first-generation chemists. These clusters are all open to new people joining them – they are not limited to women only, although women have been the primary members to date. WISC is always interested to hear about ideas for new clusters, and from people wanting to be part of WISC and to help out. The network has an inclusive approach, with the majority of resources open to anyone, regardless of gender, location, or career stage. It tries to identify issues that affect women within the community, then create resources that are open to anyone experiencing those issues. A key example of this is the parenting cluster. Caring responsibilities disproportionately fall to women, but parents of any gender are welcome in the cluster because the issues can affect them too. This was demonstrated through the pandemic.[36,38] WISC is actively researching into how teams can communicate, collaborate, and be more effective by incorporating creative and reflective processes into teamwork, and it is exploring the particular challenges that face women who are Principal Investigators (PIs) and lead teams, and how these findings might be used within public engagement. WISC is passionate about creating a sense of kinship and community, and spaces where women and those who are marginalised can share their experiences and learn from others.

Those who are involved in running WISC are all early-to-mid career scientists, with support from more senior colleagues. They have all invested time and energy in WISC because they believe that there is work to be done to ensure the retention, progression, and support of women in the field, and because they believe in the ethos WISC has taken to bring about positive change. The two factors that make the work WISC does stand out from the more general efforts to support women in science are their commitments to a field-specific approach that calls in the community, and the inclusion of a reflective, qualitative, social science approach at every step. WISC would like to think that its work might act as a model to those in other fields and disciplines who are keen to address EDI issues, and are happy to talk to and collaborate with others who would be interested to learn more.

Building community

One of WISC's aims is to build community and create a sense of kinship for women and those who are marginalised. This book is for all those who work in supramolecular chemistry, in chemistry, in science, and who need to know that they are not alone. While academia remains 'the (stereotypically) masculine and isolated place it was designed to be – a place free from children, romantic relationships, and personal problems and lives',[39(p160)] it can be intimidating for women to talk openly about their personal lives. Sue Rosser wrote: 'No one likes to feel as if they must give up their femininity, motherhood, or another characteristic they view as core to their identity in order to fit into their profession'.[4(p78)] Things have come a long way since 1988, when Traweek described the laboratory as 'a man's world'.[3(p46)] Women and those from minority backgrounds are still underrepresented, particularly at the more senior levels, making the journey towards full professorship appear daunting. For those who persevere, it can be hard to find success due to conscious, unconscious or systemic bias. It can be isolating as a member of a minority group; 'storming the tower is a lonely business, as any academic woman who has tried can tell you'.[40(p1)] Marginalisation in higher education (HE) is often thought to correlate with characteristics of the individual, such as colour, ethnicity, disability, class, and access. In terms of gender, we know that women in academia are disproportionately affected by funding structures, academic culture, research environment, and caring responsibilities. A body of work exists on academic identity and women's lived experiences as they negotiate and resist structural inequalities. However, the voices of women in STEM and their embodied stories are largely absent.

In WISC we found a group willing to trust, to innovate, and to step out of their disciplinary norms and play with qualitative research and new ideas. In doing so, the group found support and connection to understand their experiences at work, with their research teams, and to reflect on how they wanted to progress in their careers. There is an assumption within science (particularly supramolecular chemistry) that corresponding authors and team leaders are men, or women without caring responsibilities. During lockdown in the UK, one supramolecular chemist with a young child received a reviewer comment on a journal article: 'this team likely has time on their hands in the light of current in-lab restrictions'. She told us that while reading it she was literally trying to keep her daughter out of the cat food: "I feel physically sick at the thought of work right now." This book is not a collection of victim stories, it does not call out, name, and shame the senior men and institutions that perpetuate a culture of discrimination and harassment towards women. Instead it highlights the importance of community and how scientists can utilise social science methodologies for processing, sharing, and learning from experiences.

This book is uniquely positioned to set out the context for women and other marginalised groups in the field, and to share the invisible, embodied, emotional experiences of women in supramolecular chemistry, particularly as they navigate through and beyond Covid-19 and its effects on the higher education sector. It brings together findings from WISC and reflective narratives on the embodied and emotional experiences of academic work in a field dominated by white men. It is a significant contribution to work on academic identity and gender because it captures the embodied voices of women in STEM.

What's in the book

In Chapter Two we explain the reasoning and approach behind embedding a qualitative approach into our research, and why we believe it is necessary to use methods such as collaborative autoethnography and Embodied Inquiry. We discuss the meanings of these terms, explain what this equates to in terms of doing research, and how we set about creating the fictional *vignettes* that are included in the next six chapters.

Chapter Three explores the process of building an academic identity in STEM as a woman, and what is valued and rewarded in an academic career. We consider the idea of wellbeing and overwork in academia, and take a close look at how choices around motherhood can impact a scientific career. Finally, we look at options for those who choose to leave academia.

Chapter Four examines the challenges faced by women and other marginalised groups, and the lack of diversity in STEM as a whole. We consider the discrimination that results from sexism, racism, and ableism, some of the institutional drivers that are being put in place to address this, and how community and mentoring can play a part to offset the isolation and negative career impacts of being marginalised.

Chapter Five sets out the history of WISC, its ethos of 'calling in' and not 'calling out' the supramolecular community. We share the projects and initiatives we have worked on to date, including our first survey, our mentoring programme, support clusters, skills workshop, and ongoing research. We also share the publications that have resulted from our work so far, and our aspirations and plans for the future.

Chapter Six draws strongly on the collaborative autoethnography project to create stories from STEM which demonstrate a reflection of life as women principal investigators. The themes that resonated with us as a group were around overwork, self-care and community, femininity, and the emotional toll of working through this period of the COVID-19 pandemic.

Finally, Chapter Seven looks towards the future. We share findings from WISC's second survey, and show how we triangulated the data from that survey together with data from the collaborative autoethnography and the reflective work with individual research groups, to throw light on experiences of researchers and research leaders through COVID-19. We discuss the importance of community in creating change, and how WISC's work fits into wider, intersectional feminist goals. We end by thinking of the inclusive future that we would like to help create in science.

TWO

A qualitative approach: autoethnography and embodiment

WISC has created a community and place in which stories and experiences can be shared; it has given women the means and tools to do this from an embodied perspective through autoethnography and other reflexive, qualitative approaches. This chapter[*] will set out what autoethnography and embodiment are, why they are important in the context of STEM, how they are usually missing in other research, and why this is a problem. It will consider the structural barriers that are specific to STEM, and are prevalent within the culture that keeps these stories hidden.

A qualitative approach to science

Autoethnography and embodiment are not words necessarily synonymous with science. While science abounds with long words, specialist equipment, vocabulary, and acronyms

---

[*] Parts of this chapter were published in a paper for *CHEM* and are reproduced here with the kind permission of Cell Press and Elsevier.[81]

(nuclear magnetic resonance – NMR, mass spectrometry – MS, differential scanning calorimetry – DSC, or scanning electron microscopy – SEM, for example), it is less likely that you will find a group of physical scientists deliberating over a particular qualitative approach or collaborating on a piece of autoethnography from an embodied perspective. Here we will set out why and how WISC has chosen to use this approach and how it has affected the content of this book. We deliberately chose to take an approach that brought lived experiences to the fore, rather than focus on numbers. The numbers of those from marginalised genders in STEM have been known for a long time.[1–4] Some of these are set out in more detail in Chapters Three and Four. While it is helpful to know the scale of a problem before tackling it, there comes a point when as researchers we need to stop asking 'what?' and start with 'why?', 'how?', and 'why does this matter?' Like a scientist laying down a hypothesis before starting a research project, qualitative research can allow social scientists to do this too. Norman Denzin,[5(p115)] in his *The Qualitative Manifesto* of 2010, wrote a call to arms, saying: 'qualitative research scholars have an obligation to change the world, to engage in ethical work that makes a positive difference'. A qualitative approach allows us 'to improve quality of life … for the oppressed, marginalized, stigmatized and ignored … [and] to bring about healing, reconciliation and restoration'.[6(p725)] Qualitative research can be used for both justice and healing: 'a restorative view of justice that is based on indigenous ways of healing, not scapegoating and punishing offenders'.[6(p109)] Few would argue that it is time to make a positive difference, to improve quality of life, to bring justice and healing to the matter of marginalised genders in STEM. In our 2021 paper[7] we set out WISC's ethos of a field-specific approach of 'calling in' the community to support those who are marginalised rather than 'calling out' and pointing the finger at those who could do better. We reported on data from our first survey of the community (see Chapter Five). In addition, we wrote about our approach of embedding EDI expertise within scientific research, and utilising social science methods – predominantly qualitative research methods, in order to do this.

In order to write this book, we incorporate literature on women in STEM and academic identity, together with data from WISC's qualitative research projects, some of which were open to participants from all genders within the supramolecular chemistry community. Qualitative research involves a huge array of methods, approaches, and theoretical frames. In this book, and in the work undertaken with WISC, we have employed a method that uses a triangulation of data[8] between qualitative surveys,[9] collaborative autoethnographies[10] and reflective, ongoing group meetings.[11] Our approach is primarily an Embodied Inquiry,[12] which privileges knowledge that originates in the body, and the feelings, emotions, and sensations that are present. It draws on Black feminist[13] and Indigenous approaches to research.[14] This type of research has ethical implications, as it encourages honesty, emotionality, and authenticity.[15] Using data from distinct sources allows us to present a mosaic of data[16,17] or multi-layered account[13] in order to build up a picture of what is happening. Mosaics of data were first utilised when researching with young children who were not always able to articulate how they felt or the emotions they were experiencing. Multi-layered accounts allow 'data analysis [to be] a co-constructed piece of artwork'.[13(p3)] Incorporating these techniques of data capture and analysis allows us to acquire and disseminate a picture of what people are feeling and experiencing even when those emotions and feelings are not easy to put into words. It allows us to research *with* rather than *on* our community.

Creative and arts-based approaches

In order to facilitate this aim and enable us to paint a data picture, we deliberately chose to use creative and arts-based methods.[18,19] The collaborative autoethnography (CA) group were sent out 'care packages' with high-quality art materials and invited members to find items with which to play and create, and you will find images they created throughout this book. The intention was to work with research groups and at conferences in a similar way, to collaborate on collages and similar art/mark-making projects, although these plans were put on hold due to the COVID-19 pandemic. WISC set out with the aim of bringing glitter[20] and play[21] into the world of supramolecular chemistry.

The data generated with the CA group was collated along with data generated with the research groups and through the online survey to form discrete data sets. We then analysed these data sets reflectively and thematically.[22] Our objective within the data analysis exercise was to identify 'hot spots',[23] that is, to find themes that resonated with us as researchers and that were reflected in our data sets. As you read this book, you will see that in addition to using creative approaches to gathering our data, we are also using creative ways to disseminate our findings. Along with the images you will find throughout this book, we incorporate fictionalised accounts[24,25] in the form of *vignettes* drawing on the research sources. In these we have created composite accounts bringing together themes from the different aspects of our research. As discussed in Chapter One, the *vignettes* are still personal, still emotive, but they do not relate to any one particular person. This has been done to protect the researched from any perceived consequences of whistleblowing,[26] and to stop readers attempting to work out if they are included. Where we share data directly, we make this clear, though we do not attribute quotes to any individual. The *vignettes* in this chapter and in all the other chapters are fictionalised accounts. The exploring and disseminating of ideas through fiction draws on historical work by philosophers such as Jean Paul Sartre. The use of fiction and creative writing in qualitative research has been used across social science disciplines from geography to anthropology. Richard Philips and Helen Kara describe the process of creative writing within social research as the point at which 'self-reflection and social exploration connect'.[27(p10)] They go on to explain that creative writers 'write about *themselves*; so do social researchers, who increasingly bring their own stories into their work. Doing so, they reflect on their own lives and their relationships, consider what they have in common with others, and ponder how their lives and experiences converge and diverge',[27(p10)] a process akin to that used within the collaborative autoethnography that WISC undertook. In this way we can share stories that are drawn directly from lived experiences, that come from a subjective and personal perspective, and that resonate with the members of the collaborative autoethnography project, rather than objectify them, or describe them from an outside perspective. Descriptions of women in science by women (or men) who do not work in science often fail to capture the essence and reality of the challenges faced in the laboratory, or focus on external characteristics such as appearance[4] and fail to 'ring true'.

The aim of qualitative data capture and analysis is not to disseminate findings that are generalisable to larger cohorts, but to

> mentally place the reader in the physical environment where the study took place, and sophisticatedly breathe life into research participants' personal histories and commentary in ways that emotionally connect the reader to the research subject at hand ... the human experience is dynamic, and so should be our analyses in telling/writing tales of it.[13(pp1–2)]

The fictional *vignettes* created from stories shared within the autoethnographic research, reflective meetings, and the qualitative survey were written to evoke embodied responses from readers, and to illustrate themes and issues that resonated with us as researchers. At this point it might be helpful to stop and consider what autoethnography and embodiment are, and what they mean.

Autoethnography and embodiment

Autoethnography is a research approach, with an origin in the words *auto* (meaning self), *ethno* (meaning people), and *graphy* (meaning I write). Ethnography is a verb and a noun, in that one can 'do' ethnography, which in turn will result in an ethnography. Ethnography is the study of people, of social systems, and is a research approach common to anthropology and sociology. Traditionally, ethnography might include the study of remote villages and cultures, an approach much associated with anthropology.[28] Ethnography can draw on sensory experiences, such as the smells, sounds, and feelings associated with an experience or environment.[29] Autoethnography thus means the study of the self in relation to the social environment and context. This research approach, also termed critical autoethnography, is commonly used to explore subjects that are sensitive, contentious, and have personal meaning to the researcher.[30] There is no one prescribed way of conducting or disseminating an autoethnographic research study. Autoethnography is a research approach that demands a lot from a researcher. It is an inclusive approach that incorporates viewpoints and understandings of knowledge that are not limited to white, Eurocentric ideas. Validity, rigour, and repeatability, hallmarks of more traditional research approaches, are interpreted differently within autoethnographic research. Validity is gained by the researcher being critically reflexive (that is, reflecting on events, the thoughts and feelings associated with those events, and their part in creating them along with the impact and implications for those around them),[31] self-aware (that is, conscious of the thoughts, feelings, bodily reactions, and responses to events and to others),[32] and honest about their vulnerabilities, privilege, and position in the work they are doing (this is often termed *positionality* in research). This is where the idea of embodiment comes in, as it is used here to mean the totality of thoughts, feelings, emotions, sensations, images, and kinaesthetic or proprioceptive awareness that arise from the body and mind.[33] Embodied awareness allows the researcher to access data about themselves, the world around them, and how they react to others, and for this to then become part of the reflexive research process.[12] In terms of autoethnography, an embodied approach allows the researchers to notice and utilise how their body responds to the topics under consideration. An example of this might be in a conversation about laboratory space, when a woman group leader might become aware of how her shoulders rise up, the tendons in her neck tighten, and her breathing catches in her chest, feeling as though it clamps down around her heart as she describes and reflects on a battle to secure enough fumehoods for her group without coming across as strident rather than assertive. Some critics dismiss autoethnography as merely self-indulgent navel gazing, no different from biography or memoir.[34] However, autoethnography allows for diverse forms of knowledge, expression, and conception. For example, Ian Wellard deliberately uses autoethnography and calls out the 'tendency to accommodate calls for "objectivity" and ... robustness through processes which separate the researcher from the researched'.[35(p1)]

The intention behind this self-awareness is to ensure that researchers continually '[exhibit] reciprocity and vulnerability in the research process'.[13(p7)] This self-awareness is not limited to the 'data gathering' elements of an autoethnographic study, but includes the analytic process; as Evans-Winters goes on to say, this allows the researcher 'to show how data analysis can also be soul work that serves to heal thyself'.[13(p7)]

Knowledge and creativity in STEM

For all the talk on innovation in science, STEM has relatively fixed ideas on 'what counts' when it comes to research approaches and methods.[36,37] Traditionally, scientific ideas of knowledge would privilege 'evidence' over 'anecdotes' and numbers over words. This divide between quantitative and qualitative approaches to research is one that has been 'fought' within the social science arena for many decades. Traditionally, scientific knowledge is taught through the transmission of known facts, or truths, and hence a scientific approach might also be described as positivist. Positivists would prioritise empirical evidence over subjective experience – we know that this is true because we observe it to be so, and have measured it to be so. Scientific observations are traditionally objective, depersonalised, and do not take into account the personal input of the scientist making the observations. This approach to science underlies much mainstream science pedagogy, where objective knowledge is weighted with more importance than individual belief.[38] Whether it is possible to be completely objective and remove the scientist (and their background, class, cultural influences, and beliefs) from the science is questionable.[39] The positionality of the individual scientist influences the questions that they ask, the ways in which they go about answering them, and the importance they place on the patterns they see in the results – science can never be truly objective because, as it is *people* carrying out the science, it cannot happen in isolation.[40] With the use of robotic solutions and artificial intelligence (AI) increasing, there may be a future where this is not the case. However the use of AI does not necessarily mean eradicating bias.[41] Even the ideal that science 'should' be objective and depersonalised is a Westernised approach to knowledge that regards the world as an object that can be measured, understood, and explained. In contrast, Indigenous approaches take a much more holistic view of the world, allowing room for the inexplicable or ineffable.[42] When researching equality/equity, diversity and inclusion (EDI) within science, it has to be acknowledged that traditional 'scientific research can be exclusionary, while privileging formally educated, middle class, and Eurocentric styles and patterns of speech',[13(p22)] and that if we want to understand the lived experiences and recognise the impact of a lack of diversity we need to move beyond recording numbers that assess the scale of the problem, and capture the stories and subjective experiences. In order to bring about change and incorporate diversity into science, it becomes appropriate to utilise less traditional 'scientific' modes of research: 'A questioning of the objectivity of science and the purpose of science coincides with critical research paradigms'.[13(p21)] While numbers can be ordered, measured, correlated, and used to generalise findings across a population; qualitative subjective experiences cannot. As such, it may be antithetical for a traditional 'hard' scientist to employ these approaches, at least initially and until they can see and experience the benefits.

WISC and creative research

The creative research approaches that WISC uses are due to the involvement of Jennifer Leigh, the only social scientist on the Board. Jen L initially took an undergraduate degree in chemistry, and completed two and a half years as a postgraduate computational chemistry student before leaving her PhD unfinished while pregnant with her second child. She went on to complete a PhD in education. She left science, though she is a qualified secondary school science teacher and full member of the Royal Society of Chemistry. She trained as a yoga teacher,[43] and then as a somatic movement therapist and educator.[44] This training in yoga,[45–47] Authentic Movement,[48] developmental play,[49] and therapeutic technique[50] informed much of Jen L's future work.[51]

The ideas of embodiment and authenticity in research became central to her understandings of her own career,[52] academic identity,[53,54] and research.[12] She set out her path from chemistry to movement and back to chemistry again in a book chapter where she used a framework of longitudinal rhythmanalysis[55] to look at the moments of eurhythmy and arhythm[56,57] in her research career. Jen L has written extensively about her approach to using an embodied perspective and creative and arts-based methods to explore aspects of academic identity[54,58] and marginalisation within the academy.[51,59]

How does this work for WISC?

WISC's overarching aim is to address equality and diversity in the supramolecular community, and to create a sense of kinship within this same scientific interest group. Our first survey of the community allowed researchers to express the support that they wanted and share the barriers that they faced.[7] A recurring theme was the isolation and loneliness of young women, and a desire to connect with others in the field, and to see that there were role models ahead of them indicating a path to success as a supramolecular chemist. WISC's approach to surveying emphasised qualitative responses, where questions were asked in a format that allowed interpretation and extended answers. The intention was to capture the voices of the community. WISC held an open panel session at the 2019 Macrocyclic and Supramolecular Chemistry Symposium (MASC) which was attended by men and women ranging from undergraduates to PhD students and internationally renowned senior professors from around the globe. At this event we shared our website, and advertised the first WISC skills workshop, to be held in Cagliari in September 2020 (postponed to 2021 due to COVID-19). Our plan was to hold a series of panels, and host creative community spaces at the major conferences and events that would be attended by much of the global supramolecular chemistry community. We wanted to provide safer spaces where chemists could play with the ideas of reflection, make, connect, and talk to each other. These plans were upset by the COVID-19 pandemic. While there had always been a plan to have an online, virtual presence for WISC through online mentoring, resources, and the like, the more experimental and creative aspects of community building had to be put on hold. Instead, WISC decided to pivot its work towards an exploration of lived experiences of COVID-19, alongside a campaign to raise awareness of the challenges that women and those of marginalised genders faced within supramolecular chemistry (see Chapters Five and Six).

These challenges are typified by a scandal in one of the leading chemistry journals, *Angewandte Chemie*, in 2020.[60] The journal released the internet pre-publication of a paper, ostensibly on organic synthesis methods, that used language of discrimination. On publication, much of the chemistry community reacted vociferously on Twitter and other forms of social media, speaking out against the views expressed, and the very fact that it had been published. While Twitter pile-ons can be counterproductive and even harmful, the chemistry community needs to be congratulated on its actions in this case, taking up arms on social media platforms until the paper was retracted and an apology made. There is an argument that a clearly written counter-piece would have been more effective than a retraction, in order to safeguard freedom of speech and avoid concerns over 'cancel culture'. The journal's withdrawal statement reads:

> The withdrawal has been agreed as the opinions expressed in this essay do not reflect our values of fairness, trustworthiness and social awareness. It is not only our responsibility to spread trusted knowledge, but to also stand against discrimination, injustices and inequity. While diversity of opinion and thoughts can spur change and debate, this essay had no place in our journal.[61]

This demonstrates a really positive desire within the community to address social justice issues, and to decry blatant racism, sexism, and discrimination. However, the views expressed within the article were still publicly defended by a vocal minority, because it had originated from a successful senior scientist and was therefore seen as valid. The article was very quickly taken down from the journal website, many of the editorial boards resigned, and the International Advisory Board has since been reconstituted.[62]

Encouragingly, many chemistry journals responded to the paper with editorials disparaging the author's viewpoint and commenting on the need for diversity with respect to race and gender,[63–67] and these were written by chemists rather than EDI specialists. Science is not and cannot be ignorant of the increase in general awareness around social justice issues and concepts more accepted within the social sciences, such as cultural barriers, colour-blindness, meritocracy, and the impact of gender and racism. The more sensitive subject of ableism, or discrimination against those who have a disability, chronic illness, or neurodivergence within academia[68,69] is likewise not acknowledged. In Chapter Four we look at marginalisation in STEM more closely.

How WISC measures success

Success as an academic might be measured in terms of numbers of publications, citations, and amount of grant income,[70] and the individual members of WISC and WISC's Board are subject to these metrics. However, we as a collective group of individuals choose to measure success by different means. In addition to numbers engaging with WISC through Twitter, via surveys, in webinars, the Skills Workshop, support clusters, and feedback received from people engaging with WISC, we determine success by the quality of those interactions, such as the fact that senior supramolecular chemists are recommending that their PhD students and postdoctoral researchers contact us. There has also been a willingness among group leaders of all genders to engage with topics that have traditionally been ignored within science, such as combining a 'successful' academic career with motherhood in a field where there are few successful senior women role models with families, where motherhood can be seen as a barrier to progression, and where even talking about the desire to have children might be perceived as an indicator that a young woman is not serious about her career.[71] Similarly, the prevalence of sexual harassment alluded to in Chapter One is rarely brought into the open for fear of consequences. The MASC 2019 special interest symposium organised by Jennifer Hiscock, Christopher Serpell, and Aniello Palma was the first to include a no-harassment policy and anti-harassment volunteers. It was commented on positively by many young women attending, who said that they felt safe, and the idea has now been adopted by future conference organisers. The need for an upfront policy such as this speaks to the casualness of such interactions in a male-dominated field, and the ways in which young women silence themselves because of negative experiences such as implicit or explicit threats to their career for those who speak out. Those women in science who are willing to speak out about sexism and discrimination are often at the pinnacle of their career,[72,73] or have left academia[74] and thus have little to lose from raising their voices. WISC wants to create and contribute to an environment where these experiences of harassment and discrimination are the exception and not the hidden norm. Using fictional narratives in this book is a way in which to share these stories, raise awareness of the ubiquity of these issues, but without any individual paying the price for such revelations.

WISC also measures success in the way in which its field-specific mentoring programme is making an impact on those who are signing up to be mentored, and those acting as mentors. In summary, 90% of mentees expressed satisfaction with the programme and requested to continue working with their peer group for more than a year. The programme doubled in size in its first year, and has already resulted in a special issue and editorial in *ChemPlusChem* on mentoring. A second special issue in *Frontiers* on women in supramolecular chemistry has been published, and a third is in process in *Supramolecular Chemistry* highlighting first generation chemists.[75,76] Mentoring was one of the key initiatives the supramolecular community requested in WISC's first survey.[7] The importance of mentoring for women, particularly those in marginalised fields, is well recognised.[77–79]

It is absolutely critical that WISC acknowledges that the barriers faced by women and those of marginalised genders are intersectional. Intersectionality, as defined by Kimberlé Crenshaw, says that while we conduct what might be termed as feminist research and interventions, no work is truly feminist unless it includes all those who are marginalised.[80] Since the #BlackLivesMatter protests of 2020/21 there has been an upsurge of attention given to the lack of racial diversity in science, with hashtags on Twitter bringing together communities around #BlackInTheAcademy, #BlackInStem, and #BlackInChem among others. Much work is needed to diversify and decolonise curricula within science, and this is part of the effort necessary to diversify the scientific body and faculty.

To sum up, WISC measures success quantitatively and qualitatively. We want to demonstrate through numbers how our approach is impacting on individuals and the community within supramolecular chemistry, and the change we are instigating. We also want to reach out and connect emotionally in order to share the reasons why change is necessary, and the difference that it makes when it happens. We want to combine rational claims with the emotional impact that they have on people's lived experience (see Figure 2.1).

Figure 2.1: Which path should I take?

First vignette Maria, 32, Early career

I'd consider myself fairly successful in my career. You know, I have a permanent position, and my group are doing well. We're not a huge group, really quite small but the things we are doing are working and I have had a couple of grants awarded recently. It's funny, I thought I'd be happy when they came through but I was actually quite scared! It's a good job I am prioritising my work right now. My partner is supportive, he understands that work is my focus, and that means that we actually live about two hours away from each other at the moment. At one point we were in different countries! It can be lonely, but it does leave me a lot of time to work late and at weekends when I need to. These new grants mean that we'll be expanding pretty quickly with a new post-doc and two new PhD students. I do feel the pressure in that though – this thing that you've been wanting and working for suddenly comes good and now you have to deliver! I really don't want to get it wrong hiring this post-doc. What really helped was being able to stop and reflect and to think about it as well as to talk to others. I couldn't believe it when I found that I was not the only one who had had these thoughts and fears! I even heard stories about getting the hiring all wrong and we chatted about the different things that you can do to see if someone was suitable or would fit into the group. I know getting this right will set the tone for my group for the next few years. I really want to create a culture where everyone feels valued. I've been a post-doc myself enough and worked alongside others enough to know that some people out there might be good on paper but can just rub others up the wrong way.

Sometimes it seems cultural, you know? But I shouldn't say that. I know I need to be really aware of bias when it comes to hiring. I've asked my friends if they have any masters students that might be interested in the PhD positions, and I've also asked around if there are any good potential post-docs but I know that I want to be open to anyone who comes. Having to do everything all online because of COVID makes it harder though!

One thing they suggested was doing a task-based activity, and asking for a one-page research idea or fellowship outline. That should help to see if they can write. I need to get a LOT of publications out of these grants so I need people who can write. But I also need people who can get on…

It's going to be interesting to see if I get a lot of women applying. I know that sometimes women do. I know a couple of friends have groups that are nearly all women; that's just the way it turned out because of who applied. I've never had a PI who was a woman myself. I don't know how much difference it makes.

If you'd told me this time last year that I'd have been looking forward to those meetings and that I would have been getting out my notebook to draw stuff I'd have laughed at you! Reflecting and drawing aren't things we really do in science. But it has really helped me figure out how I feel about things, and untangle all the emotions about it. I want to do a good job, and I think this experience is helping me to do a better one. We've had lots of hard and personal conversations – conversations I think everyone should have about privilege and equity and racism – but I have got so much support just from finding a community like WISC.

THREE

Building academic identity in the context of STEM

This chapter contextualises what it is to build an academic identity as a woman or other marginalised gender in STEM, and the particular challenges faced by these individuals. In the chemical sciences, the progression and retention of marginalised groups including women is an issue, and the barriers they face are intersectional. We consider academic identity – that is, what it means to be an academic and to succeed in academia – and the pressures faced by early-career academics in general. We discuss the concept of wellbeing in the context of acceleration and overwork in neoliberal academia before turning to the particular barriers faced by women and mothers (and those with caring responsibilities). Finally, we explore leaving academia, and options open to those with a science PhD.

Academic identity

Being an academic is something that is often bound up with a fair amount of mysticism and misperception. In movies and books, academia and campus life can be portrayed as some kind of halcyon idyll, with young people willing to learn, experiment, and have fun under the benevolent guidance of wiser men (and they very often are assumed to be men – a Google image search of 'what does a professor look like' displayed 29 white men in the first 22 images returned by this search in June 2021). However, in more recent times this has changed, with academics portrayed as 'the bookish but socially challenged swot or the egomaniac self-publicist that communicates his or her elevated status at every available opportunity'.[1(p68)] Les Back goes on to say:

> Academics themselves don't much like other academics, and often feel a deep estrangement from their colleagues as people. Perhaps part of the problem is that our forms of self-presentation are tied to the modern academic desire to be taken seriously – that is, the embodiment of entrepreneurialism, 'being smart' and 'world-class' braininess.

Back, along with other commentators and researchers, blames the shifts seen in the 'modern academic' on the neoliberal university, which captures academics' dispositions towards hard work and achievement and overlays them with demands. Back states that 'Academics should see themselves first as teachers'.[1(p46)] but this sentiment is at odds with advice given to early-career academics to focus instead on their research, grant income, and markers of esteem in order to attain success.[2] Teaching, while an integral component of the kinds of professional and educational development courses found for early-career researchers in the UK, Australia, and parts of Europe, is not recognised or rewarded in the same way as conventional research outputs or cold hard cash.[3]

The current accelerated pace of the academic world is sometimes termed 'fast academia'.[4] Work is experienced like a treadmill,[5] with the associated symptoms of overwork and illness endemic.[6,7] Universities have to run as businesses, with the purpose of higher education changing from individual transformation to producing goods and workers for the knowledge economy.[8] Squeezing budgets to maximise profit results in a need for academics to increase the speed and productivity of their work without losing quality – though, as Bourdieu says, 'you can't think when you're in a hurry'.[9(p28)] Speed and acceleration of this kind is tied to the political economy of capitalism.[4]

Many universities are including career maps as part of their promotion and progression pathways, laying out indicators and criteria that are expected to be met at certain stages on the road to seniority. The goal at the end for many is the elusive full-time permanent contract, and the title of professor as their reward for producing 'world-class ground-breaking research', achieving excellence across all aspects of their administration, service, teaching, and research, with the hierarchy weighted heavily towards the prestige of research. Steps along the way, post-PhD, might include securing a position as a postdoctoral researcher, then, after a faculty position, promotion from Assistant Professor or Lecturer to Associate Professor (securing tenure in the US) or Senior Lecturer/Reader. Neoliberal academia has led to the casualisation of higher education, with post-docs as well as adjunct or zero-hours faculty hired to cover teaching and administrative duties on precarious contracts creating an excessively bottom-heavy pyramid structure. In addition, there are 'far fewer women than men at the top of the academic hierarchy; they are paid less and are much less likely to have had children'.[10(p3)] Navigating this pathway (map notwithstanding) can be incredibly overwhelming for an early-career academic (see Figure 3.1).

While much has been written on academic identity and how this is achieved and supported, generally it must be recognised that the pathway and requirements in STEM are different from social sciences or arts and humanities. There are commonalities, such as the need to author publications, bring in research money, and build a reputation, but the leap to becoming an

Figure 3.1: Being an early career researcher is like fighting a giant squid

independent researcher (from post-doc), together with running a research group and building a lab, are unique to science. One of the easiest ways to conceptualise this is perhaps when considering research leave or sabbatical. Sabbatical is often used within the social sciences and humanities to take time out of the office, away from teaching and administration duties, in order to write a monograph. These long texts or books are seen as necessary to achieve promotion and progression. While they may have less kudos in research assessment exercises than peer-reviewed journal articles, they remain the lingua franca for a successful academic. In science, however, books or monographs beyond textbooks (which are generally not recognised as research) are virtually unknown. For a scientist to progress in research, they often need to spend *more* time in the lab, and *more* time with their postgraduate students. Although there are benefits to sabbatical trips such as visiting other labs or institutions to form new networks and collaborations or learn new skills, being away from their own campus day-to-day would not progress their career or research as they need to be present for their lab groups. Unfortunately, this feeds into the perception that in order to achieve in science, the best thing to do is commit to overwork – that is, to work all the hours into evenings, weekends, and forgo any work–life balance at all. Life becomes a round of chasing ideas, becoming a pathway more openly accessible to those with the time available to dedicate to the job, with good support networks, and to those without additional caring responsibilities.

Wellbeing in academia

Academia is often seen as more than just a job; instead, it becomes part of a person's identity, spilling over into every aspect of their life,[11] which results in overwork. In the context of overwork, academics' wellbeing is often forgotten.[12] However, there has been a proliferation of 'self-help' books aimed at academics, entreating them to gain balance,[13] be happy in their chosen career,[14] or advising them how to leave academia successfully.[15] Such books are aimed at all those involved in academia, regardless of gender, contract type, or marginalised status. Wellbeing as an academic may seem something of a holy grail. When Jen L told the collaborative autoethnography group that she had bought a book called *How to be a Happy Academic*,[14] the response from the rest was derisive laughter that there could ever be such a thing, since a happy academic still would not have time to write such a book, and who else would know? Everyone in the group has chosen to be an academic, and would like to be happy, but it will take more than a 'how-to' book to accomplish this.*

In a study exploring how academics with an embodied practice (such as yoga, running, meditation, and the like) expressed the tension between a non-judgemental ethos and their academic work, which was critical, cognitive, and competitive, Jen L described wellbeing in the context of academic overwork:

> Wellbeing is a funny concept, apart from a lack of consensus over how it is spelt, there are many discourses over what it actually means. The UK National Account of Well-being[16] defines it as a dynamic thing, a sense of vitality that people need to undertake meaningful activities, to help them feel autonomous and as if they can cope. However, as Richard Bailey puts it, 'many of these discussions take it for granted that wellbeing equates to mental health'.[17(p795)] In turn, mental health seems to be conflated with being 'happy', or with factors that are personal, and to do with whether life is going well for the individual or not. Griffin[18] explicitly connects wellbeing with happiness, similar to Aristotle's idea of it

---

* The deliberate inclusion of this anecdote discloses the 'messy talk of reality' that aims to use autoethnography pull down barriers between the private and public as a transgressive stance.[64(p178)]

being the fulfilment of human nature.[19] Philosophically, wellbeing can be associated with either a hedonistic 'desire fulfilment', whereby it is achieved when an individual has sated their desires, or as a more objective theory which judges whether things are good for people or not.[20] This latter view is one which sometimes results in lists of factors that indicate wellbeing or quality of life[21] and quantitative measures of wellbeing.[22] However, quality of life should be seen as a dimension of wellbeing rather than be conflated as the same thing.[23] Practices that increase awareness and the quality of consciousness have been reliably shown to have a significant role in increasing wellbeing.[24] Embodied practices such as yoga, mindfulness, and Authentic Movement, a structured dance form that draws on Jungian principles,[25] contribute to wellbeing through enhancing this sense of present awareness and a wholeness of mind, body, and spirit.[26] Wellbeing is often measured quantitatively, and yet if we are looking for embodied answers to research questions, how should we go about collecting data?[12(pp224–225)]

The answer to this, as far as this book is concerned, as described in Chapter Two, and as will be shown in Chapters Six and Seven, is to triangulate or create a mosaic of data from different sources – including that from collaborative autoethnography, images, fiction as research (in the form of *vignettes*), surveys, and ethnography – so that we can capture and share lived experiences, increase the visibility of challenges, such as those detailed previously within this chapter – for example, the spread of work into life and lack of balance – and find ways to ameliorate or banish them.

Women and mothers in academia

Achieving success in academia or 'storming the tower' is 'a lonely business, as any woman in academia who has tried can tell you. Even with the support of other feminists in one's own institution or country the sense of isolation can be overwhelming'.[27(p1)] We should note here, of course, that feminists do not have to be women. However, more women in academia are on casual, fixed-term, or precarious contracts.[28] The pathway to seniority is not always clear,[10] particularly in a context where admitting struggles undermines professorial identity,[29] and where minorities (including women) are also battling other barriers. The rules of academia can seem opaque:

> a commonly used informal description of the upper echelons of academia is a network of 'old boys clubs' ... [which] remain highly effective for their members, exclusionary to women, and ... play a tacit role in recruitment and selection, and the furtherance of [some] men's academic careers to the detriment of their women counterparts.[30(p91)]

Some men still believe that 'cutting edge science and engineering remain out of reach of the vast majority if not all women'.[31(p49)]

In science, women's experience of academia is often mediated by their experience of motherhood or perceptions of what it might be like to be a mother (or not) in academia, and when they might fit in having a baby[32] (see *vignette* this chapter and Figure 3.2). Motherhood in academia can be seen as a fleshy contrast to the 'academy's "floating head" syndrome; how people are expected to function as disembodied brains, not connected to bodies or families outside of academic pursuits'.[33(pp51–52)] Babies and children are the antithesis of 'floating heads', full as they are of milk, snot, vomit, sh*t, and very obvious and present visceral needs. The tension between motherhood and academic success is felt across every discipline.[34–36] Mary Ann Mason describes children as 'a wonder and a blessing, not a problem; but motherhood is. Child rearing does not occur in a vacuum; decisions about

Figure 3.2: The constant question of how to time family into the career motorway before the stork crashes into the brick wall of age

motherhood are bound up with societal expectations, the nature of the workplace (and how it works for or against mothers), and women's personal needs during various life stages'.[37(p7)] This is not helped by the fact that the age at which the majority of women have to focus on establishing their post-PhD academic careers occur when they are often in their mid-to-late 20s to 30s, meaning 'the career clock and the biological clock are on a collision course'.[37(pxvii)] Indeed, some women advise others to intentionally put their careers on hold while they have a young family: 'If I could give you a gift, it would be the patience to recognize that childhood is precious and fleeting and that science will be waiting for you with some awesome mysteries when your children become adults'.[38(p66)] Women often defer having children until later in life, as 'they fear that they will not be taken seriously'"[37(p15)] if they have a child as a student or postgraduate. Success in academia is 'measured by productivity, not time, and assistant professors tend to put in long hours in order to meet departmental expectations ... Mothers are less likely to travel to attend conferences to present their research findings to other scholars – a critical step in career advancement'.[37(p37)]

In science, the idea of advancing your career, of being seen to be successful and productive is key:

> The scientist who toils away in the lab, tied to her lab bench at all hours, skipping meals and hunkering down to finish just one more grant application, is not pregnant. She is not running out the door at 2:30 pm to pick up the kids. In fact, to do so would be considered disloyal and unscientific in the patriarchal culture of academia.[29(p77)]

Academia, just like much of society, is patriarchal,[39] and this is particularly evident in science:[40]

> many successful male scientists have a multitasking primary caregiver wife who tends to carry the domestic load, or at the other end of the spectrum, there is the male scientist whose partner or spouse is an academician and typically a few steps behind him on the career path. In contrast, the highly successful female scientists advance their careers within a very small spectrum. Either their partner/spouse works full-time outside the home and typically holds a high powered job or they are single.[41(p99)]

In 1995, 66% of women scientists and engineers were married to male scientists and engineers, and women remain more likely than men to have academic partners.[42] The scarcity of senior women in science who have had successful relationships and families is a stark reminder to young women that historically they would have to make a choice between their career and a being a mother.[40] Mason and Ekman write:

mothers who do persist do remarkably well. They don't do as well as men, but they compete favourably with women who don't have children ... The most successful take parental leave following childbirth for a few weeks or a few months and then return as full-time workers ... Somehow these mothers overcome the emotional and physical pull of the infant and the forbidding judgement of a society which increasingly sends the message that mothers who can afford to stay at home should do so.[37(p53)]

The need to return to work quickly after childbirth may be due to fear that 'in competitive fields, a person who takes time off from work may be "scooped" and miss out on, or at least delay, a chance for career advancement'.[43(p104)] In the collaborative autoethnography group, women PIs shared experiences of having to respond to reviewers' comments on papers and grants while caring for a new-born baby. Academic work of this kind cannot easily be passed to someone else to complete: 'balancing career with family, particularly at the time of childbirth, is perceived to jeopardize the careers of women scientists and engineers more than any other single factor'.[42(p43)] The expectation that in order to succeed a mother has to 'overcome' the natural pull towards their own child and minimise the time they spend with them while young for the sake of the career is disturbing and discriminatory: 'nearly half the female scientists in the US leave full-time science after their first child is born. In comparison, 80 percent of male postdocs and female postdocs without children stay in science'.[44(p209)]

The struggle of the working mother is nothing new.[45] Aviva Brecher, who described herself as an 'over-sixty baby boomer scientist', in 2008[46(p25)] wrote of her fellow women scientists: 'it is astonishing that they are asking today the same questions about how to successfully manage and blend careers in science with the demands of motherhood and family life that we struggled to solve thirty years ago'. The assumption that women have a 'second shift' at home after work, comprising the bulk of the childcare, and the emotional labour of running a family (if not the physical duties on top), is also not new. For many scientists, the idea of returning to work part-time is unrealistic, as the aspect of work that gets reduced is research rather than the more time-consuming duties of administration, service, or teaching. In order to succeed, they have to find a way to do it all. Balancing a career with a family slows down the careers of women scientists, but not those of male scientists.[42] This is not to say that the skills they learn as a working mother are not valuable; A. Pia Abola writes:

> I know that the time I spent at home caring for my children has made me a much better scientist. I am more efficient with my time and better at planning and prioritizing; I am more pragmatic and goal-orientated; I am humbler and better at dealing with overcoming my own shortcomings and those of others; and I am better at negotiation and compromise; and I am much better able to tolerate the tedium and myriad little failures that accompany work at the bench.[47(pp125–126)]

The COVID-19 pandemic has brought new challenges to women in academia. In addition to the broader struggles around scarcity of research funding,[48] increased redundancies in the sector,[49] and the emotional impact of living through a pandemic,[50,51] additional burdens have been placed on women and mothers. Early data suggested that there was an impact on women's ability to publish, with comparative rates declining when compared to those of men.[52] WISC's second survey showed that for the majority of PhD students or post-docs without caring responsibilities lockdown was a productive time, although this was not true for those responsible for research groups[53] (see Chapter Seven). Some male academics with children also found this period productive (see, for example, one man who wrote a book in six weeks while his wife took on responsibility for their children and home).[54] For many women, however, the coronavirus outbreak was a time that led to decreased opportunities for work, increased load of home duties (such as schooling children[55]), and increased emotional labour.[56] At the time of writing, work on women's experiences of mothering through COVID-19 is beginning to appear in print. Mothering as a journey is not uncomplicated, as for many it brings together the challenges of being a woman, the expectations of society, and the responsibility of caregiving.[36] Mothers share a host of personal reflections and short chapters edited by Andrea O'Reilly and Fiona Joy Green in *Mothers, Mothering and COVID-19: Dispatches from a Pandemic*. The book highlights the '"third shift" – the emotional and intellectual labour of motherwork',[57(p20)] and gives evidence to support the claim that the pandemic has had a devastating effect on gender equality. In a report from Kris 'Fire' Kovarovic, Michelle Dixon, Kirsten Hall, and Nicole Westmarland *The Impact of COVID-19 on Mothers Working in UK Higher Education Institutions*,[58] they looked to explore the impact and experiences of mothers of children under 18 working in UK higher education and share examples of good practice across the sector. Using a combination of interviews and an online survey, they found that, unsurprisingly, mothers took on the burden of responsibility for childcare, and that this had a negative impact on their physical and mental health. Mothers had less time for self-care. In contrast, the majority said that their workload had increased, but the quality of their work had been affected, primarily due to combining paid work with childcare. Over 80% said that it was impossible to work uninterrupted from home. Over half said that this had already negatively impacted their career progression through missed opportunities. A similar combination of pressures was felt by the mothers in our own research (see Figure 3.3). Kovarovic and colleagues[58] found that expectations of 'business as usual', poor communication, too many changes, and uneven use of furlough or leave, along with too many meetings, lack of suitable home-working equipment, and hiring and/or pay freezes were unhelpful responses from

Figure 3.3: Carrying it all

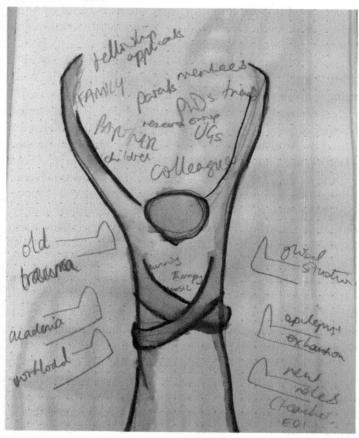

institutions. In contrast, initiatives such as: giving teaching or marking relief; cash payments to say 'thank you' for work during the pandemic; bold, decisive decisions communicated clearly (such as cancelling all first-year students' exams); and carers' funds to help with additional childcare costs were all given as examples of best practice that led to staff feeling valued. Mothers who had been recipients of initiatives like these said that they felt as though their wellbeing was prioritised, they were allowed to do their jobs, and this best practice helped mitigate fear of future impact on their careers.

Leaving academia

Gender balance and marginalisation in academia is not the same across the academic disciplines. This is in part due to scientific approaches, assumptions around gender, and the constitution of concepts such as rigour and validity, which are embedded from school teaching and texts. It is 'old news' that not a single woman scientist was named in the UK 2020 single science GCSE syllabus. According to the Royal Society of Chemistry, the chemical sciences have a particular issue with retention and progression of women, and stated in 2019 (pre-pandemic and any gender-based concerns that lockdown has raised) that at the current rate of change we will never reach gender parity. If the positions of power within science are predominantly white and male, this 'sends a message to our undergraduate and graduate students, half of whom are female.".[37(p107)] In higher education science, where textbooks are often used on reading lists, and are authored by senior (white, male) scientists, achieving a gender balance (and decolonising) science is not easily achieved. The nature of learning and knowledge and the types of skills is particular to STEM, and, as discussed in Chapter Two, rarely includes reflective practices. This, in turn, impacts on how women and marginalised groups construct their identity, and process and disseminate their experiences. Further, this lack of diversity among scientific leaders 'may allow unintentional, undetected flaws to bias ... research.".[42(p182)] It is well known that women and those who are marginalised leave the sciences. This problem is broadly known as a leaky pipeline[59] – a term that in itself is problematic[60] – but how can we fix this and keep women in science?

One challenge is the dominance of the perception that if you have a PhD, then you 'should' aspire to remain in academia. This is reinforced by attitudes that might be vocalised to colleagues, such as 'if they're not going to stay then you might as well give up trying to get them to produce work of a decent standard'.** To an aspiring academic this attitude reinforces the belief that if you cannot find a job then it is down to your failure, your inability to work hard enough, or simply not being good enough. The reality is that the vast majority of those who graduate with a PhD do not remain in academia. In the UK, HEPI (the Higher Education Policy Institute) reported that 67% of PhD students aspired to continue working in academia, but only 30% were still doing so three and a half years after graduation.[61] In the sciences, an even smaller fraction continue on to full-time permanent positions, as many leave during the 'post-doc years'.[62] Career options for science doctoral graduates can include working in industry, in professions allied to industry, law, patent attorney, finance, teaching, editing (for example, for specialist scientific journals), science communication, research support (for universities or funding bodies), among others. However, for many, leaving academia can feel like failure, or so they are told.[63] The disconnect between industry and academia is in part due to the lack of 'employability' preparation that many PhD students receive in their training.[61] Professors can be oblivious of other options for their students, or unwilling to see them, and either cannot prepare them because they have not seen life outside academia themselves, or fail to prepare them for the reality of other work because they do not value it. Either way, students can be left unclear on how they might actively utilise the skills they have acquired.[15] One woman who had recently left academia to pursue a career in industry, leading a drug development programme, drew an image to express how she felt (Figure 3.4). She described how she had been pruned into a certain shape driven by the needs and demands of academia, only to find

---

** This was said to one of the authorial team by a colleague, and represents a prevailing attitude.

Figure 3.4: My ideas being 'trimmed' to only keep the few viable and profitable ones

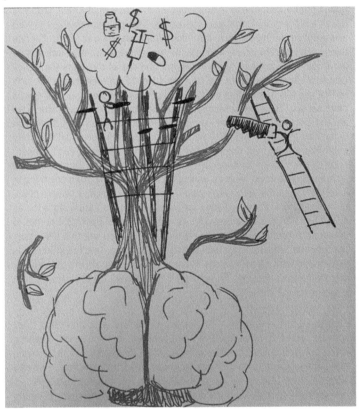

that her new industrial bosses had other motivators (profit). While EDI policies and practices may be enforced differently outside of academia, the barriers against women and other minorities do not simply disappear. The next chapter will look in detail at the impact of discrimination in science. The number of women continuing to senior positions is still small compared to the proportion of those graduating, as is also the case for those with other protected EDI characteristics such as race or disability.

Second vignette Adi, 23, International PhD student

I don't think I want to stay in academia. I don't know what I want to do yet; there's a big part of me that wants to use my degree as I have worked so hard and spent so much time on it, but I don't think academia is for me. For one thing I don't see any faculty who look like me – there aren't any Black women in my department. I look at my supervisor and see all the hours she works. Last year I know she was in the lab until 11 or 12pm every night. She tells us that she has to work that much just to get things done. She's not even a professor yet. There never seems to be any time to take stock of where we are and everything we've done – as soon as something works we just move on to the next thing. Before COVID we used to celebrate in the pub if the group published a paper, but these days it just doesn't happen. It can't happen. I don't think it's only my supervisor either. I remember how burnt out all my lecturers looked when I was an undergrad. Always rushing from one thing to another. It's not that they didn't help – they did or I wouldn't be here now – but I don't think it's what I want for myself. Someone I know got a job as a lecturer straight after a post-doc. Almost unheard of right? But he has to do so much teaching and with everything online I can see that it's almost breaking him. At least my group isn't as bad as some. One friend I know is expected to work 11 hours a day 6 days a week and her supervisor regularly sets meetings at 8pm on a Friday night. I don't think I could cope with that; I want to have a life!

Even doing this is stressful. I cannot tell you the levels of stress that I just seem to cope with on a day-to-day basis and see as normal now! Worrying about not getting enough data, worrying when things don't work, worrying that I am not living up to the huge sacrifices I made to come here. So much worry and fear. Leaving my family and boyfriend and only seeing them a few times a year is horrible. I worry that when things aren't working I'm letting my supervisor down. I have to live up to all their expectations of me and make this worthwhile. My whole support network is in another country. There are wellbeing services and things at uni – but they are really general and the waiting lists are so long it's like I'll graduate before I get any real help. My mental health is definitely suffering. As an international student the worry about the financial and emotional cost of my PhD is huge. And every time I want to apply for any kind of financial help they ask for a million things and really intrusive information so I just give up.

Even if I did want to stay, it's not as though getting a job is easy. There are so few positions out there, and everywhere people are being made redundant.

I heard somewhere that a CV that won't get you into a post-doc these days would have been better than one that would have got you tenure 30 years ago! It's as though expectations just keep going up and up and up and the pressure goes up with them.

I also know I want to have a family and I just don't see how that's possible in academia. None of the really senior women I see in the field have children. I know that there are more younger ones coming through who have had kids or who are having kids, but who knows if they will make it to professor? There is so much holding women back in the field, doesn't being a mom just make it harder? How could I be the kind of parent I want to be and work that hard? That said, I don't know what I am going to do. Industry seems to be almost as bad. Maybe publishing or editing is the way forward. Something where I can use my science but also have a life.

FOUR

Challenges for women and marginalised groups

This chapter looks at the experiences of academic work from an intersectional feminist perspective, including the context of traditional gender boundaries. These are often compounded for women in science as they take on or are asked to provide additional equality, diversity, and inclusion support either formally or informally through student support and the lack of recognition for this work. We consider marginalisation in academia, and the casual sexism and harassment that minorities may face. The global COVID-19 pandemic has exacerbated pressure for many academics. In looking at data that includes undergraduate numbers, postgraduate numbers, and progression through the postdoctoral system and on to senior roles internationally for those who are marginalised, we consider aspects that contribute to progression, such as bias in publishing, citations, funders, and processes such as networking, winning funding, and securing tenure (in the US). We also consider how these might be affected by intersecting barriers such as gender, disability, and race.

Lack of diversity in science and chemistry

It is not news that there is a lack of diversity in STEM. Across the world there are inequalities, and systemic and structural racism are endemic.[1,2] Although the focus in this book is on gender, these things are always intersectional. When race and other factors such as religion intersect, barriers compound, and the onus is often on those who experience those barriers to educate others about their impact.[3,4] Within academia it is no different: 'we operate in the teeth of a system for which racism and sexism are primary, established, and necessary props of profit'.[5(p27)] Certain bodies in academia are deemed out of place, they do not belong, and as a result they get stuck.[6] Diversity across academia is still something to be celebrated rather than taken as a matter of course.[7] In science, 'minoritized people of colour who participate in STEM are positioned as replacements in the mostly White STEM professoriate',[8(p1)] and in the US racial diversity is lacking across every level of the education system and workforce.[8] The proportion of Black professors in the US engineering and computing faculty has remained constant at around 2.5% over a ten-year period.[9,10] In 2014 the proportion of Black Americans in the population was 12.4%, but less than 4% of all undergraduate and postgraduate students graduated in engineering.[8] More recent data is available from the National Science Foundation, and it is clear that underrepresentation remains an issue.[11] To address this lack of diversity, and to stop a system whereby students of colour are encouraged to enter a space that has systemic barriers against them, it is necessary to have 'researchers who specialise in racism, sexism, and other forms of bias to be part of the discussion and search for solutions'.[8(p5)] In this way, it will reduce the likelihood that structural barriers and individual privilege will be ignored due to science being purely objective and meritocratic:

> Meritocratic perspectives suggest that sociocultural norms in science education are rooted in the 'impersonal

characteristics of science'[12(p269)] and produce objective sociocultural standards for communication of knowledge. Such perspectives align with positivist productions of scientific knowledge within value-neutral environments, positioning concepts of racialized or gendered microaggressions as subjective forms of preferential treatment. This value-neutral ideology protects inherited advantages, creates insider/outsider dynamics, and necessitates forms of cultural capital. Among students from traditionally marginalized populations, failure is viewed as an individual consequence rather than a reflection of systemic oppression.[13(pp19–20)]

Science is not as objective or meritocratic as it claims to be. Scientists

> have a more difficult time than other kinds of workers do in perceiving themselves as discriminatory ... science has a vested interest in the idea of the intellectual meritocracy. It is important to scientists to believe that they act rationally, that they do not distort or ignore evidence, that neither their work nor their profession is seriously influenced by politics, ambition, or prejudice.[14(p59)]

There are still far fewer Black academics in UK academia than would be expected.[15] While 3% of the population identify as Black,[16] as of January 2020 there were only 155 Black professors out of 23,000.[17] Only 10 Black scientists were funded by UKRI (UK Research and Innovation) in 2020 and the only Black chemistry professor was not among them.[18] White STEM academics are three times more likely to become full professors than their Black counterparts.[19] In the wake of the 2020 Black Lives Matter protests attention was drawn to inequality and lack of diversity across science.[20,21] It is known that Black academics and other minority groups are judged less fairly in student evaluations, which are often used as the basis for promotion and progression criteria.[22]

If we turn to look more specifically at chemistry, it is easy to see that across the chemical sciences there have been calls for action to bring more diversity into the workforce – see work by Ackerman-Biegasiewicz and colleagues,[23] Menon,[24] Reisman and colleagues,[25] Urbina-Banca and colleagues,[26] and the RSC,[27] among others. There are particular issues that need to be addressed around racism. For example, within chemistry journals,[28] and the lack of Black, Asian, and minority ethnic graduate students.[29] These issues have been known for some time, and in the UK and US national subject-specific societies, such as the Royal Society for Chemistry[27] and the American Chemical Society[30] have put together concrete recommendations to address them.

In this book, we have taken a feminist perspective; however, without intersectionality there is no feminism. Intersectionality is a term first coined by Kimberlé Crenshaw[31] to describe the multiple barriers of sexism and racism faced by Black women. Intersectionality has since been co-opted to include other instances of compounding factors faced by individuals who experience intersecting marginalisation due to being Black, Indigenous, or a person of colour, having a chronic illness or disability, class, religion, sexuality, ethnic origin, and the like. Within academia, the region of the world in which a researcher is based might also marginalise them and their research. Women of colour (WOC) face a 'double bind' of racism and sexism 'the environments in which WOC STEM faculty must work are often not ideal … these suboptimal environments often lead to faculty discrimination, intentional attrition (e.g. choosing to leave for a variety of personal or professional reasons), and unintentional attrition (e.g. not earning tenure)'.[32(p56)] Women of colour are often victims of tokenisation (where they are differentiated from their counterparts in unfair ways, on display, expected to conform, be socially invisible, stereotyped, and lack sponsorship), pioneerism (that is, being the first minority in the department having to serve as the first or only representative of their gender or race), marginalisation (where their contributions are overlooked, ignored, or minimised), and microaggressions (frequent intentional or unintentional derogatory comments).[32] Science is not as democratic and meritocratic as it could be, though hopefully times are moving on from when 'democratisation ... applied itself only sparingly to people of colour and, to a significant degree, all women in science remain unreal to the men with whom they work'.[14(p54)] We must acknowledge that change in a world full of tenured positions is slow by default. If the efforts of the people trying to change things are not acknowledged properly or at all, they will become frustrated and stop altogether. This is the last thing we want.

Disability in academia and science

Being disabled, neurodivergent, or having a chronic illness in academia is not the norm.[33] Up to 30% of the general population is thought to have a condition that would be recognised under the 2010 Equality Act,[34] compared to 16% of the working age population, and just 4% of academics.[35] While disclosure rates are slowly increasing across the sector, this varies according to discipline,[36] with the physical sciences and subjects with the greatest gender imbalance having the lowest disclosure rates.[37] Ableism in academia is endemic.[38,39] Decisions to disclose a condition or disability are personal,[40] and have to take into account a weighing up of perceived and actual risks versus benefits. Such decisions may factor in the particular condition that an individual has, and how it is perceived in society,[41] as well as more general stereotypes of disabled people as scroungers, workshy, or lazy.[42] For example, cancer or multiple sclerosis may be perceived as a more *worthy* condition than a contested illness such as fibromyalgia, or mental health issues. Similarly, there may be internalised ableism or preconceptions about neurodivergence such as autism or ADHD (Attention Deficit Hyperactivity Disorder).

CHALLENGES FOR WOMEN AND MARGINALISED GROUPS

There may be motivating and demotivating factors from the institutional perspective, such as promotion and progression processes, performance indicators, and the like. In addition, there may be political reasons an individual might have to 'step up' as a role model for colleagues, students, and society.[43]

Disabled people face many barriers and microaggressions,[44] and these continue throughout education[45] and in society.[46] Within academia, these discriminations can be from external funding bodies[47] as well as internally within institutions, and the culmination of such discrimination and barriers is the absence of academics with disabilities, chronic illnesses, or neurodivergences.[37] In part, this can be explained by a predominance of the medical model of disability within science, which describes a disability as a deficit of the individual.[48] In contrast, the social model of disability[49] would instead label the environment or society as disabling. For example, an individual in a wheelchair is only disabled when the ramps and lifts are not in place to let them get where they need and want to go. The medical model allows able-bodied and neurotypical people to see those with disabilities as 'lesser', and less human. It is common for those who are 'out' about their disability, neurodivergence, or chronic illness, or who are a minority due to their religion, sexuality, gender, or other protected characteristic, to become involved in work that supports equality, diversity, and inclusion. However, this work is not always recognised in formal progression processes by the institutions,[6] and can in fact rebound negatively on the individuals involved.[50]

Gender imbalance in science and chemistry

Women have historically been thought of as lesser beings,[51] from Aristotle, who wrote on women's general intellectual inferiority,[52] to Galen, who described female embryos as polluted.[53] In the Victorian period, prevailing biological and medical ideas reinforced the inferior status of women, making it more difficult for them to pursue a career in science.[54] Unfortunately, this attitude persisted well into the 20th (if not 21st) century. While interviewing an eminent male scientist for her book *Women in Science*, Vivian Gornick recounted: 'He confided in me that it was simply a matter of the nervous system ... "Women may go into science, and they will do well enough, but they will never do great science"'.[14(p26)] This is despite assertions from male scientists that they do not have unconscious bias towards women: 'We don't do that ... Science is objective. We only hire the best, and we know it when we see it'.[50(p184)] An example of such bias is how Ellen Daniell was denied tenure as the first woman in her department, despite having positive reviews from her students and recently funded grants in which she was described as having 'a good publication record ... positively cited, and appears well on the way to becoming the authority on adenovirus chromatin'.[55(p58)] This experience led to Daniell first leaving academia to work in industry, and later leaving science altogether. Women are often penalised with worse performance or teaching ratings than men.[50]

There are many even more disturbing stories from women who work or worked in science. For example, in her second year of postdoctoral research, Sue Rosser became pregnant with her second child. On telling her principal investigator, he "told me to get an abortion because the pregnancy came at the wrong time in the research ... we needed to gather data intensively over the next several months in preparation for renewal of the grant'.[56(p41)] These comments made Rosser feel that becoming pregnant had jeopardised her career, resonating with the discussion on motherhood and science in Chapter Three, and Xie and Shauman's[57] findings that having a family slows down the advancement of women in science. Nancy Hopkins related a story to Rita Colwell of an incident as an undergraduate student when Francis Crick visited his friend James Watson, Hopkins' mentor. Crick 'zoomed across the room, stood behind me, put his hands on my breasts and said "what are you working on".[50(p63)] Colwell, the first woman to lead the National Science Foundation in the US, commented: 'women knew that men looked at them as sex objects; that was just life'.[50(p63)] Colwell herself was told 'girls don't do chemistry'.[50(p7)]

The gender gap is wider within STEM disciplines than in many others. This is not news. However, science is not a monolith, and there are variations across disciplines and countries. For example, from 2014 to 2019 in the UK, there were more women than men enrolled to study medicine and dentistry, subjects allied to medicine (including nursing), biological sciences, veterinary science, agriculture and related subjects, and architecture, building, and planning.[58] However, the reverse was true in physical sciences, mathematical sciences, computer sciences, engineering, and technology.[58] The figures are similar for the US, with 50.5% of all science degrees awarded to women in 2012.[59] This makes statements such as 'in 2019 in the UK 35% of STEM students in higher education were women'"[60] difficult to substantiate, as it is not clear which subjects are included, and the granularity that shows that women are choosing to enrol in some subjects and not others is absent. Even data for the 'physical sciences', which combines chemistry and physics, can be misleading, as in the UK chemistry attracts a higher proportion of women undergraduates than physics.[61] However, as discussed in Chapter Three, even with high numbers of enrolments at an undergraduate level, it is another thing for women to continue on to find careers and success within science wherever they are in the world.[62] Women academics are subject to the same pressures to achieve citations as men. However, across academia women are cited less than men,[63] and this effect is particularly evident in the physical sciences, where it was one of the only subject groupings that showed evidence of systematic bias against women in a large study conducted by Elsevier.[64] When it comes to peer review, there is evidence to show that harsh reviews, which are antithetical to the collegiate purpose of improving scientific research[65] can be moderated when the reviews are published.[66] However, to date there has been little work that seeks to specifically address unconscious bias towards women and other minority groups in the peer reviewing process.

In chemistry in the UK, historically women owe thanks to individuals such as Ida Smedley (1877–1944) and Martha Whiteley (1866–1956), who 'both pursued outstanding but very different scientific careers whilst endeavouring to improve conditions for women, and after a protracted battle, they were among the first group of women to be allowed membership of the London Chemical Society'.[54(p169)] These women also initiated a dining society that met three times a year, forming for themselves the kind of community and network advocated by many senior women in the sciences as a way to combat the isolation faced by minorities.[50,55,56] In Chapter Five we will return to the importance of community, and the role that WISC plays to support the retention and progression of women. For now, we note how the pressure of historical and present-day barriers for women and marginalised groups can lead to the feeling that they are being crushed by the weight of what has gone before (see Figure 4.1). In the chemical sciences, the lack of retention and progression for women and all those with protected equality and diversity (EDI) characteristics is pronounced.[27] This is highlighted by data – for example in the UK in physics the percentage of women choosing to study at A-level at school or college is around 25%, with the proportion of women reaching full professor approximately 9%.[67] However, the percentage of women choosing to study chemistry at undergraduate level is over 45%, while the proportion of women reaching full professor is still only 9%. As we have seen already, more women are employed on short-term precarious contracts.[67] Women author fewer papers and are cited less.[68] Proportionately fewer women sit on editorial boards, are nominated for awards, and far fewer file patent applications.[65] As discussed, gender is of course not the only

Figure 4.1: Crushed by chemistry

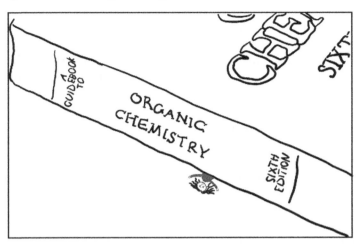

factor that those in STEM face, and when this is looked at intersectionally, these barriers compound, contributing to a much larger overall effect.

Institutional drivers to promote equality and inclusivity

There are institutional initiatives and drivers to promote quality and inclusivity, which include programmes such as the US POWRE (Professional Opportunities for Women in Research and Education) from the National Science Foundation. Evaluating the programme after it had been in place for a decade, Sue Rosser found that women still experienced continued barriers around balancing career and family, their low numbers, stereotypes held by others, the context of tight resources, which particularly impacted women, and overt discrimination and harassment.[59] Lessons have been learned from the US ADVANCE programme.[69] This programme to address organisational change for gender equity in STEM directed $130 million to encouraging institutions to evolve policies and practices that supported women. In the UK, the Athena SWAN (Scientific Women's Academic Network) is a national scheme that promotes and certifies gender equality. It was initially managed by the Equality Challenge Unit, which is now part of AdvanceHE. Athena SWAN was originally set up with the principle that in order to address gender inequalities, there needed to be commitment and action from people at all levels of an institution. It had the aim to tackle gender imbalance in science, unrepresentative cultures and attitudes, and to do this through addressing the personal and structural obstacles women face.[70] The initiative has since expanded to encompass all disciplines and include an institutional level of certification, but critical questions remain as to whether it has in fact reduced sexism in science or academia, and whether Athena SWAN has become less about *doing* gender equality, and more about *certifying* gender equality.[70] Similar initiatives exist for promoting inclusivity in UK academia, including the Race Equality Charter, Disability Charter, Stonewall, and the like. However, as with Athena SWAN, there needs to be a continual process of questioning and refining the impact they have on policies, procedures, and cultures. These drivers need to be examined critically, as Campion and Clark did for the Race Equality Charter, so that ideals are translated into actions.[71] Campion and Clark wrote:

> the REC is not perceived as a significant vehicle for progressing race equality work in award-holding institutions. Rather, it is mostly applied as an enhancement tool to help shape and sustain existing race equality initiatives that produce incremental change. This, we argue, suggests the REC's intention to inspire race equality approaches that favour institutional strategic planning at the highest level, is yet to be realised.[71(p18)]

At the level of individual funders and professional bodies, there is also ongoing work in this area. For example, the Royal Society for Chemistry has collated data on bias and discrimination against women and is putting into place an action plan, as it has done to support chemists who are LGBT+, and is working to support those with disabilities.[27] But there is little literature around the experiences of those who choose to stay and who want to progress in the field. This means there is a need for a publication dealing with the particular context they face, and providing narratives of their subjective, lived experiences. Within science generally, and specifically within chemistry, there is a focus on numbers, rather than lived experience. This book, and the work of WISC more generally (see Chapter Five) is looking to address this, by creating a sense of community, kinship, and a programme of subject-specific support and mentoring.

Community and mentoring

When Colwell was setting out a list of things to reform in science, she stated:

> we do not need to cater to women in science. We need only give women an equal chance to achieve. The best of 100 percent of the population will always be better than the best of 50 percent of the population. Once all the talent in our country can compete on a level playing field, decisions about who to hire and who to support can be made on the basis of brains and ability, not gender, ethnicity, or national origin.[50(p194)]

Colwell said that key ways women and girls could support themselves to have a successful career in science were to form or join study groups that include other women, to go online to meet kindred souls and meet with them bi-weekly so as not to be alone, and to find a mentor. Rosser states that women need to find role models and mentors to support their journey.[56] Similarly, Daniell uses the model of support she received from her Every Other Thursday group, and suggests that 'regular meetings, with the objective of advice and support in a confidential context, is critical to groups and other mentoring relationships'.[55(p174)] They are not alone. Mary Ann Mason advocates that women find themselves a mentor, although she acknowledges that 'mentors are not easy to find, in part because it is not usually in anyone's job description. As women rise in their careers they must make sure they bring younger women along with them and take responsibility for setting up a mentoring program in their workplace'.[72(p117)] Mason goes on to say that support networks are critical in order to recruit and retain women in science: 'continuous practical and emotional support is needed, and is a key to women's success in research and academic science'.[72(p119)] Across much of Sue Rosser's work on women in science and STEM she has been consistent with her message pleading for the structures and systems to change and do more to help women, as much as for women to come together in order to advocate for themselves and find ways to become a critical mass and be present in the spheres from which they have traditionally been absent in science.[56,59,73,74] Similarly, Emily Yarrow stresses the importance of community networks for women in academia,[75] and Mahat and colleagues include finding a trusted mentor in their strategies for thriving in academia.[76] The message seems clear, and is supported by many others who write from their own experiences as women in a marginalised environment:[77,78] Find your community, meet with them regularly, and share what you are going through in order to access practical and emotional support in a safe space.

These safe, or 'safer', spaces are where women and those who are marginalised can 'let off steam', reflect on, and process their experiences. This might include a structured method by which they allocate time to each individual and the 'work' they wish to do,[55] action learning sets,[79] peer mentoring, or a more informal routine. Such communities, and places to share experiences without judgement, have been particularly vital during the 2020/21 COVID-19 pandemic. Throughout lockdown the ordinary, everyday pressures of academia and managing a work–life balance were exacerbated for many due to the additional load of homeschooling for those with children at home, pivoting to online learning, and the emotional burden of worry and fear for friends, families, and research groups (see Figure 4.2). In Chapter Six we share stories and experiences from our own regular meetings through the COVID-19 pandemic.

In the next chapter we set out the story of how WISC began, and the model that we have for building community in supramolecular chemistry. We share how we have set up a programme for mentoring, support clusters, and, in Chapter Six, how the regular collaborative autoethnographic meetings have allowed us to communicate transparently, express what we have experienced, find practical and emotional support, and collaborate professionally.

Figure 4.2: COVID-19 'Can you mind the baby?' 'Buried and burned'

Third vignette Paula, 33, Early career

Until two months ago I was the only woman in my department. There doesn't ever seem to have been more than one woman here at a time, I'm not entirely sure why. I'd like to think that it isn't because they just want to have a 'token' but it does sometimes feel like that. You know, one woman, one person of colour. Actually, I *wish* there was one person of colour. It's not great optics that our students are diverse but the faculty looks all white. I get asked to sit on a lot of committees. Hiring committees, ones where the department needs to be represented. I never like to say no because I know that the department will have to decide in the next couple of years whether they are going to make me permanent, and promote me. I need to have good relationships with them so that they ask me to collaborate on projects with them. If I say no then they might say that I haven't given enough service to the department, or I haven't been collegial enough, and I might miss out on those opportunities. Now that I think about it there haven't been that many opportunities coming my way from the department – most of the collaborative work I do has been as a result of my own networks outside the university.

I do worry though that it all takes me away from my research. I don't see work of this kind getting recognition when it comes to grant applications or promotion. It doesn't seem to matter as much as papers and funding. Some of my friends told me that I should just say no more often, and put myself first. Instead, I have been getting up at 5am just so that I have a couple of hours to write. I haven't really published as much as I would like to yet, I'm only beginning in my career and developing my group. I don't want to become known as 'that woman' if I keep banging on about it being unfair. I saw that happen to a friend when I was a PhD student. He was in a wheelchair and had to fight for everything. He wasn't even lab-based so you'd think it would be easy enough to get a computer with the right keyboard, and for him to be able to get into the postgraduate office, but still. He became a bit of an advocate for disability rights, and I think that is one reason why he couldn't get a post-doc position. He left science in the end; I don't know what he's up to now.

I find that a lot of our women students want to work with me, or talk to me. They haven't met many other women in the field, and they want my input or advice. They see me as a role model, which is lovely, but it also takes away time from my own research ...

Still, at least I haven't experienced any direct discrimination! Some of the things I've heard, particularly at conferences! Conferences are good when there are other women about; it can feel a bit intimidating when you're on your own. These days I have a little group of other women that I tend to stick with. Some of the things they've told me ... Senior professors asking them what underwear they have on, being told that they shouldn't be in the lab because their hands are too small to hold solvent bottles, or their wombs will sink because women aren't meant to stand up at a fume hood! It didn't seem to occur to that prof that women also have to stand up if they stay in the kitchen ... I don't think they are trying to be sexist, they just don't realise that we face a barrage of little comments and innuendos all the time and it gets tiring.

FIVE

WISC: Women in Supramolecular Chemistry

This chapter focuses in on the field of supramolecular chemistry. It gives an overview of the kind of interdisciplinary STEM research it encompasses, as well as the history and background to WISC.

It was Jean-Marie Lehn who first coined the term 'Supramolecular' and, alongside Charles Pederson and Donald Cram, won the 1987 Nobel Prize in Chemistry for innovations establishing this scientific field. Lehn described supramolecular chemistry as 'chemistry beyond the molecule' with the aim of 'developing highly complex chemical systems from components interacting by non-covalent intermolecular forces'.[1] Therefore, supramolecular chemistry can simply be defined as the study of the non-covalent interactions between molecules. This is often referred to as a host–guest scenario in which reversible interactions temporally hold together two or more chemical species, thus forming a supermolecule. In essence, you can think of these supramolecular reversible interactions that hold molecular species together as a type of atomic hook and loop fastening, similar to those developed by VELCRO®. If you pull the hook and loop partner strips apart, they remain unaltered; bring them back together and they will stick to each other, holding two pieces of material in place. This is a process that can be repeated, depending on the quality of your hooks and loops, an infinite number of times, and is the key principle behind all areas of supramolecular chemistry.

However, despite the scale and scientific diversity within this area of study, there is a significant lack of representation for women. For example:

- Only 12 publications were included in a 2018 Special Issue of *Supramolecular Chemistry* highlighting the achievements of women across the international community. This can be compared to the 27 articles included within a 2013 Special Issue dedicated to Professor Rocco Ungaro.[2]
- Only 5 international women principal investigators attended the 2018 UK RSC's annual Supramolecular and Macrocyclic Chemistry (MASC) interest group symposium (150 attendees).
- Male speakers outnumbered women 4:1 at the 2019 International Supramolecular and Macrocyclic Chemistry (ISMSC) conference.
- Only 2 women have won the RSC's Bob Hay lectureship prize for supramolecular chemistry in the period 1991–2021. Similarly, in that time only 2 women have won the Izatt–Christenen Award, and only 1 woman has won the Cram Lehn Pedersen prize 2011–2021.[3]
- Until 2019 only 2 women had served on the RSC's MASC committee since 2001, and in 2021 2 out of 25 were women, and all committee members presented as white.

As we saw in Chapters Three and Four, the chemical sciences are a discipline that has had less success at achieving parity or representation when it comes to gender balance and diversity. When women are in a minority, then it is more likely that they will feel isolated. One way to combat feelings of isolation is for women to establish their own networks to help themselves and each other. This is how WISC began.

History of WISC

WISC developed along a number of different threads that coalesced at the same time. The first, and probably the most important thread, was the friendship between a small group of young, early- and mid-career women in the field: Jennifer Hiscock, Anna McConnell, Cally Haynes, and Claudia Caltagirone. They realised that they needed more support and organised themselves to have bi-weekly online meetings to generally connect, talk about their research, publications, and grant proposals. These women had something that earlier generations of women in science did not have – a peer group. Their friendship laid the ground for what became an informal online peer-mentoring group. They read each other's grant and fellowship applications, collaborated on projects, and supported each other through job applications and moves to different countries. By 2019, they began to be approached by other women in the field who wished to join their unintentional peer-mentoring group, including Marion Kieffer, and realised that the successes they were seeing as a result of their support of each other was something that could be of benefit to others. They felt that the reason their peer group was so valuable was because it was a field-specific approach to mentoring rather than general or even discipline-specific. They garnered the support of two of the most senior women in supramolecular chemistry – Professors Kate Jolliffe and Michaele Hardie – to support their idea to roll out clusters of mentoring to other women. However, before they rolled anything out, they wanted to survey the supramolecular community to ascertain whether their own ideas about supporting early-career researchers were the same as those in the rest of the community.

The second thread comes in at this stage. As they had little experience of designing and implementing surveys or any form of social science research, Jennifer Hiscock reached out to another friend, Jennifer Leigh. The Jens had met while Jen H took a mandatory teacher training programme as a requirement for her probation as a university lecturer. Like many of the scientists forced to take this programme, Jen H was a challenging student, who would much have preferred to spend her time on research, as opposed to being locked in a room with a bunch of social scientists. Jen L was the unfortunate teacher of the class, and quickly realised that if she failed to win Jen H round then the whole group could easily become disrupted. Luckily, she was able to draw on her own background in science[4] and the pair formed a connection. When asked by Jen H to have a look at the questions that the group wanted to put to the supramolecular community, Jen L first gave a quick lecture on the need to write survey questions that would give the answers to the research questions, then offered to write the survey herself and put it through ethical approval processes in her department. Jen H responded by co-opting her onto the WISC Board. Jen L brought rigour and knowledge of traditional social science research methods such as surveys and the like to the group. As a higher education researcher she also brought her expertise in academic identity and marginalisation in the academy, and, more controversially, a focus on embodied and creative research approaches designed to reveal and capture the unspoken stories that exist in society.[5] The methods WISC has utilised are described in Chapter Two.

WISC believed that mentoring would be the most valuable thing to offer initially. From their own experiences they were particularly aware of the 'jump' from post-doc to independent researcher, but they were cautious of projecting their own experiences and assumptions onto others. WISC set about designing a logo and initial website with Rosa Burton (*Burton Designs*) and put out its first survey. This garnered 100 qualitative responses within the year it was open, reaching data saturation long before this.[6] At this time these initial WISC members began reaching out to other members of the supramolecular community to help support WISC's growth. This included Emily Draper and Anna Slater, followed by Davita Watkins, Nathalie Busschaert, and Kristin Hutchins, who also now hold official Advisory Board positions within the WISC network.

In addition to the small group mentoring programme led by Marion Kieffer, and as a direct response to the results of this first survey, WISC has set up community clusters. The first of these, spearheaded by Emily Draper, was the Parenting Cluster. This included all parents, of any gender, whether they were biological parents, step-parents, foster parents, adoptive parents, or prospective parents. WISC has since formed two new clusters: for those with a disability/chronic illness/neurodivergence and for first-generation supramolecular chemists. These clusters act as arenas for targeted support and discussion that have arisen from survey responses and emails to WISC, and they all approach this targeted support slightly differently. The Parenting Cluster is very research-focused, and aims to provide support, a place to share experiences, to learn, and to document its findings. The Disability/Chronic Illness/Neurodivergence Cluster arose out of Jen L's work on ableism in academia,[7–10] and is supported by Anna Slater. This group meets regularly to provide a safe space for people within the field to share experiences, give support, and learn from others. The cluster supports much of the advocacy work that is carried out by the NADSN STEMM (National Association for Staff Networks' Science, Technology, Engineering, Mathematics and Medicine) Action Group and members.[11,12] In 2021, four members of the cluster, Kira Hilton, Orielia Egambaram, Jen L, and Anna Slater, won funding from the Royal Society of Chemistry for a project to imagine the future accessible laboratory.

The third cluster, for first-generation supramolecular chemists, was launched September 2021, and will likely be much more structured in terms of its activities and research. Half of the WISC board identify as first generation, and it has been recognised that those who are new to higher education experience more barriers than those who already have 'capital' and understand the ways in which higher education and career progression work in academia.[13] Capital is anything that confers value to its owner, and is used here to represent the idea that some people have knowledge, or connections that enable them to gain knowledge about the way things work.[14,15] Within academia and higher education, this might include pre-university understandings of which combinations of subjects are required for study at university, which universities are determined as 'good', and, as a graduate student and early-career researcher, access to the 'hidden handbook' of academia that lays out what is needed in order to gain funding, publications, promotion, and progression. Capital does not necessarily mean that you know the information yourself; rather that you know the right questions and the right people to ask. The 1st Gen Cluster is likely to encapsulate a large proportion of people from ethnic minorities and diverse backgrounds, including refugees. WISC aims to support equality and diversity within the supramolecular community and hopes to give tools to others by acting as a model for how this framework can 'call in' the community to support its own in other fields.

The final threads came as a result of WISC forming officially. WISC initially secured finances through the Royal Society of Chemistry Diversity and Inclusion fund to develop the network, and started to put together grant applications for other funding streams, several of which were successful (see later in the chapter). WISC now has a formal structure, a website (thanks to the Biochemical Society and Dave Robson), and a programme of events and ongoing funded research work, including a part-time research assistant in Sarah Koops, funded by Christian-Albrechts-Universität zu Kiel (Germany). Within a year of its official launch in November 2019, WISC had a paper on EDI accepted by one of the most prestigious international chemistry journals.[16] The speed of the network's growth has surprised us all. It speaks to the need to do something different to address marginalisation and inequality in science, and a willingness to try and play with new ideas.

WISC initially set out to address equality, diversity, and inclusion issues in supramolecular chemistry, and to support the retention and progression of women in the field post-PhD. It quickly evolved into an international network of women supporting these issues in the field, not limited just to gender, and to 'calling in' the community to support its own. It is easy enough to point the finger at science and scientists, at the structures that support academia and industry, and to call them out as being sexist, racist, and discriminatory.[17–19] In Chapters Three and Four we briefly touched on the fact that calling out, complaining, or 'whistleblowing' as a member of a marginalised group can have adverse career repercussions.[20–22] As a young woman in science, much as you do not want to be harassed, you also do not want to become known as someone who complains or who causes trouble. As we saw in Chapter One, the low rates of successful prosecution for sexual harassment combined with the minimal consequences for senior men in science who have been accused of the same[23–25] mean that the impetus is often going to be to 'put up and shut up' rather than to make a fuss. WISC's approach of 'calling in' rather than 'calling out' aims to create an atmosphere where the community can talk about and address its problems and challenges, and work to find solutions.[26] We have found an encouraging amount of support for this approach from senior members of the supramolecular community in person and on social media. One tweeted in vocal support of the reach of our first paper 'WISC are putting my generation to shame'.[27]

'Calling in' and 'calling out'

The concept of 'calling out' on matters of racism, sexism, ableism, and the like is commonly known.* Within STEM,

---

* This section is based on an article that appeared in *Chemistry World*.[34] Parts have been reproduced and built upon with kind permission from the Royal Society of Chemistry.

we can all think of instances where members of the scientific community have been called out on inappropriate behaviour or for their use of inappropriate language. The act of calling out is a direct challenge to another. As such, it can be an intimidating thing to do, since standing up to someone senior to you and telling them that what they are doing is not okay can be frightening and requires a lot of emotional labour. Emotional labour means managing or regulating the feelings and emotions you may have around a task in order to express yourself in a way that means that the task gets done. The canonical example of emotional labour is flight attendants who are required to remain smiling and pleasant in the face of the most stressful of situations (or cantankerous passengers). For the one calling out, there may be unintended consequences to their career, as they might then be seen as a troublemaker. The converse is that being called out can be a threatening thing to be on the end of, particularly if you do not understand the transgression that you have been accused of, or you were doing your best to be supportive of EDI issues but did not get it completely right and are being called out as a backlash. Fear of a backlash might inhibit someone from even trying to make a change or statement on an issue related to EDI. Additionally, responses to being called out are quite often defensive, and can result in behaviour, arguments, or actions that become even more hurtful to those who are doing the calling out.

In contrast, 'calling in' has a different ethos. Rather than pointing the finger at others, it is an invitation to discuss something that might be uncomfortable in a safe environment, without fear of getting it wrong, and then to pull together to make positive changes. The field of supramolecular chemistry has an active group of senior researchers who saw the need to develop a field-specific community and welcoming environment for early-career researchers, eventually forming MASC (Macrocyclic and

Supramolecular Chemistry Royal Society of Chemistry interest group[28]). WISC began by reaching out to this community through the two annual MASC symposia; first conducting an anonymous survey to ascertain their views of the problems facing women and those who are marginalised, and then to ask for support in addressing the issues by acting as mentors to those at an earlier career stage. This approach was deliberately non-confrontational, and as such was an invitation to be part of change that will benefit the entire community. To date, unlike some other advocates for EDI within STEM, WISC has had overwhelming support for its activities, and nothing but encouragement from those most senior in the field. We want the work we are engaged in to bring about actions and inspire change from others. In 1979, Audre Lorde, a self-proclaimed Black feminist, poet, and warrior, said that if we want to change things, we need to do them differently.[29] Although scientists are not always willing to try things that challenge their assumptions of concepts such as rigour and validity, WISC has taken a creative and reflective approach to ongoing research projects to humanise the reasons why equality work is imperative. While WISC aspires to be an 'agent of change' or, to use the phrase Sarah Franklin[30(p158)] coined, 'a wench in the works'. We want to do this in an effective way that does not put the careers of our board or members at risk. Even so, standing up and putting our heads above the parapet in this way can be terrifying. One member of the group wrote about how her heart felt when she intentionally put herself out in this way:

> I am so aware of my heart pumping and expanding; not sure whether terror or pride mostly terror and fear and instantly on to the next thing but where is the
>
>   p   a   u   s   e
> the rhythm
> the beat.

Projects

To date, WISC has secured funding and support for different projects through professional/learned bodies including the Royal Society for Chemistry, the Biochemical Society; national funding bodies such as UKRI, and the Royal Society/BA APEX;** institutions including the University of Kent, Universita degli Studi di Cagliari, and Christian-Albrechts-Universität zu Kiel, and companies including Scot Chem, ChemPlusChem, Crystal Growth & Design, and STREM Chemicals.

These aspects of work interweave and are interconnected (see Figure 5.1) and are all dedicated to meeting the wider WISC aims of supporting women and those who are marginalised to progress within supramolecular chemistry through creating a sense of community and kinship. We summarise some of the main activities we are engaged in in the following sections: the website and logo; events; surveys; support clusters and mentoring; research; and publications.

Website and logo

One of the first things WISC did was to create a website. WISC's logo and website uses colours reminiscent of the suffragettes (see Figure 5.2 for the design of the 2021 pin badge for the first WISC skills workshop and future events, which incorporates the WISC logo). While WISC would never condone militaristic action, we recognise that if we

---

** 'In partnership with the British Academy, the Royal Academy of Engineering and the Royal Society ('the Academies') and with generous support from the Leverhulme Trust, the APEX award (Academies Partnership in Supporting Excellence in Cross-disciplinary research award) scheme offers established independent researchers, with a strong track record in their respective area, an exciting opportunity to pursue genuine interdisciplinary and curiosity-driven research to benefit wider society'.[38]

Figure 5.1: Interweaving threads of WISC projects

want change to happen we must be responsible for bringing it about rather than waiting for someone to do it for us. We knew that having a logo and identity was really important to establish a presence within the sector. The website acts as a repository for resources, including papers and links to current EDI work in chemistry. There is information on events, such as webinars, and WISC's own international chemistry skills workshop. There are short descriptions of the Board and Advisory Board members, our terms of reference, links and information on the support clusters, and links to any active surveys or research projects. From the website, people can sign up to the mentoring programme as a mentor or mentee, and contact us to enquire about activities. Although we have had relatively regular conversations about the need or desire for WISC merchandise (notably glitter-infused hoodies as well as more practical display boards and signs), the COVID-19 lockdown and cancellation of in-person events meant that these aspirations were put on the backburner.

Figure 5.2: Design for enamel badge incorporating the WISC logo

Events

After an initial in-person panel event at the early-career researcher MASC event held in the summer of 2019, WISC launched in-person at the December 2019 MASC symposium, which was attended by supramolecular chemists of all genders and career stages. The COVID-19 pandemic put paid to plans to run similar invited sessions at MASC and ISMSC (the International Symposium of Macrocyclic and Supramolecular Chemistry) and to hold the first WISC Skills Workshop in Cagliari in September 2020. However, the pivot to online facilitated a collaboration between WISC and vMASC (the virtual MASC early career group), which resulted in a successful series of webinars. These have included a WISC panel, a session on science communication with Vivienne Parry, work–life balance with Professor Jeremy Sanders, careers sessions focusing on options outside of academia, sessions on intellectual property and gaining research funding, and a session showcasing the work of mentors and mentees.

The first WISC Skills Workshop, led by Claudia Caltagirone, was postponed to September 2021, and was shifted to become a hybrid event, with vMASC running the virtual elements. The workshop focused on providing a gender-balanced programme of speakers, opportunities for early-career researchers to present, and retained the area-specific focus that WISC specialises in while championing equality. The virtual element, which was free to all who registered, allowed participation from across different continents and countries, including Africa, Asia, Europe, and North and South America. The workshop report[31] and a conference report published in *Nature Chemistry*[32] show that it met its aims. The plan is for the skills workshop to be a bi-annual event.

Surveys

Our first survey looked to explore what the supramolecular chemistry community wanted from an organisation like WISC.*** In one year, we had 100 responses, from PhD students through to senior independent researchers. Of these respondents, 81% identified as women, and 10%

*** The data shared here was published in *Angewandte Chemie*[16] and is shared with their kind permission.

identified as having a protected characteristic (such as race, religion, ethnicity, LGBTQ+, disability). The largest group of respondents were independent researchers, followed by PhD students and postdoctoral researchers. The majority of respondents were positive about WISC, and what we were trying to do:

> This is a wonderful initiative and I would be absolutely delighted to contribute to it at any capacity!

> There is a shortage of role models for women. I have participated in conferences with 1% of women as speakers, it is clear that women are underrepresented. We need to support each other through experience sharing and make each other visible, at least among ourselves, to be able to promote each other.

> Continue with the great ideas! It would be good to discuss issues surrounding the effects of parental leave on careers and how to minimise this, particularly with respect to getting funding bodies on board with these adaptations.

There was support for all the ideas we put forward for events, webinars, research, and mentoring. The importance of mentoring was mentioned:

> From personal experience I have found that regular mentoring and support makes a world of difference in terms of career development. As a female in science, there are often other factors to consider, such as family constraints, and having support from other women who are in a similar circumstance may be advantageous.

> Advice, knowing their route into supramolecular chemistry, mentoring sounds really useful too :-)

The findings from the first survey led to the development of WISC's initial offerings to the community, including a mentoring programme that used a model where small groups of mentees were paired with a mentor at least one career stage ahead of them. Resources on mentoring were included on the website, and mentees and mentors were asked to complete a mentoring agreement, and to meet once a month. The mentoring groups are now regularly surveyed to check satisfaction, and to date 90% are very satisfied with the programme and would recommend it to others.

The survey also asked about barriers to developing as an independent researcher, and career breaks. We found that there was a divide in how people spoke about career breaks according to gender – the men who responded shared that their experiences had been positive:

> Half a year parental leave. No troubles returning, at least that I noticed.

> My return was smooth, as this was managed within my contract as a faculty member.

> Once I returned, all of my former colleagues and peers were amazingly supportive.

Whereas women often had very different experiences:

> Support is patchy, expectations are wildly different, and I lost a first authorship which may have affected how people perceive my career. Some in the community are incredibly supportive. Some less so.

> I started as a professor immediately after finishing mat [sic] leave, so I didn't have work to catch up. Overall, my transition was smooth. However, reviewers DON'T SEEM TO PAY ATTENTION to the 'Leave of Absence'

entry in the CV. on my grant review, only 1/3 reviewers acknowledge the leave.

On returning I found I was behind on my research and unsupported.

Younger women were very apprehensive about taking career breaks in the future:

I am quite scared about having kids before getting permanent/more stable.

From past hearsay and departments I've worked in, it's almost as though women who have taken career breaks in chemistry seem to just fall off the radar, with no support from the department, and that those that really push forwards with their career are seen to be really 'pushy' or 'over-reaching', which is awful! Needs to change.

Findings from WISC's second survey, exploring experiences of being in and running research groups through COVID-19 and the impact on mental health were submitted to a leading chemistry journal, and are shared in more detail in Chapter Seven.

Support clusters and mentoring

As mentioned, the support clusters and mentoring programme have been running since early 2020. The clusters all have regular events, or activities, and the mentoring programme had 7 mentors and 19 mentees from three different continents as of July 2021. The mentoring groups were constructed to bring 1–4 mentees together around a more senior academic (one or two career stages ahead) in order to create a support network beyond the usual one-to-one mentoring relationship.

The groups usually meet once a month, with more frequent informal catch-ups between the mentees encouraged. Mentees are given the opportunity to reflect on the sessions through a mentoring log. The doubling of numbers between the first and second year of the mentoring programme shows the increased interest of the community for field-specific mentoring. We expect the network to continue to expand, with the usual intake of new mentees increasing after conferences and events where WISC talks about the network and programme. Such events should be facilitated again in the post-COVID-19 world.

Research

WISC has a number of ongoing research projects, and the methods used within these are discussed in Chapter Two. These projects include a collaborative autoethnography project that crosses continents to enable participants to find points of connection as women PIs explore life inside and outside the laboratory. Findings from this are shared in Chapter Seven, and were used to feed into the *vignettes* woven throughout this book. The collaborative autoethnography was originally planned in order to explore how women PIs could enhance the communication of their teams, facilitate more moments of inspiration and creativity that occur in order to increase the quantity and quality of their scientific outputs.

Linked to this, Jen L has also worked with two research groups, using reflective and creative approaches directly to enhance their capabilities as scientists. The groups met with her bi-weekly, and each meeting had an aim, or topic for discussion. She enjoys this work, describing herself as "filled with fizz" after a session. Topics have included the qualities and attributes of a chemist, what pressure feels like, and motivations, for example. The work from the groups fed back into the collaborative autoethnography sessions, which then fed into the research groups of all the members.

WISC has a public engagement project running, linked to the BA APEX-funded study. This project specifically looks to engage with people without a science degree, and is aimed at young girls and those from marginalised groups. Outputs utilise materials and footage from the main study, and edit them into YouTube and YouTube360 video shorts. It will also incorporate in-person workshops using some of the creative and reflective tasks employed with the laboratory research groups, when such events are allowed again after COVID-19 restrictions are eased.

Publications

To date, WISC board members have published a number of blogs, an editorial,[33] an article for the UK professional magazine,[34] two peer-reviewed papers,[16,35] and have more in the pipeline, including a chapter in a book titled *Women in Academia: Voicing Narratives of Gendered Experiences in Higher Education*.[36] We wanted to ensure that our work was disseminated to a wide audience – of those who work in and around STEM, as well as those who work in and around academia.

Aspirations and plans

WISC has aspirations to increase its reach into other continents. Jens L and H were asked to contribute a workshop to a 'bootcamp' run by EFeMS (Encouraging Female Minds in STEM[37]) aimed at young Black women in Africa, encouraging more women to progress into STEM careers. The workshop drew on lessons learned from the public engagement project, taking from creative SciComm (science communication) techniques to explore how these young women experienced the barriers and challenges of being Black in science. The YouTube platform was then used to shape the footage generated in the workshop to form part of the public engagement content, thus increasing the visibility of Black women in science and reaching a wider audience. The RSC funded an extension of this collaboration, where Black ambassadors will be brought to the UK (COVID-19 allowing) in spring 2022, to spend time in research groups led by women. Depending on the career stage and interests of these ambassadors, they will either focus on SciComm elements, or be engaged doing actual supramolecular chemistry research. The University of Kent is funding virtual elements of the bootcamp and creative workshop, as well as a documentary on the experiences and journey of the ambassadors.

Our second survey highlighted the need to bring WISC's successful model to other areas of the world. Nathalie Busschaert, Davita Watkins, and Kristin Hutchins are spearheading a US version of MASC, and will lead the second WISC Skills Workshop. We are keen to explore links already made with India, and to respond to and support WISC members who have made contact and who have responded to our surveys and attended events. We are aware that our virtual and online presence enables participation from women and those who are marginalised across the globe, but our focus has been concentrated on Europe and the UK as that is where the majority of the Board work and live. While we aspire to extend the reach of WISC and engage more people in our events and projects, we also measure success by looking at how our work affects people on a personal level as much as wider work in the field.

Above all, we want to hold true to our original aims – to support the retention and progression of women in supramolecular chemistry through building an inclusive community and sense of kinship. We feel that the model we have created – using an area-specific focus, utilising EDI expertise, and bringing qualitative research approaches to scientists – is something that can be replicated across other countries, subject areas, and disciplines.

Fourth vignette Mira, 27, Post-doc

When I heard about WISC I was really keen to join a mentoring group for women. I've been offered mentoring before, but this was a bit different. The way they set it up was to have a small group of us all more or less the same career stage, and then to have a more senior mentor. All of us are in the same field, which means that we can be really specific about what we are aiming for, who we need to talk to, if we need advice when something isn't happening, or if we need to look for funding or a job. I've always heard of the old boys' network but this is a bit like a new girls' network! Just not so exclusionary! I really liked that most things WISC do are open to everyone. They've made a big thing about being inclusive to those who are trans, and they have special support clusters for parents and for people with disabilities. I haven't used any of them yet, but I recommended one to a friend and I think she's been in touch.

It's nice to feel that I am part of a community. It's hard making friends and getting to know people when you're on short-term contracts and moving about every two years. The precariousness of the whole thing gets me down. I guess I'm lucky I don't have a partner because it means I'm 'free' to chase after new jobs no matter what country they are in and don't have to worry about them having to follow me and find a job or deal with a long-distance relationship. My friends who have moved overseas for jobs say it's the hardest thing. I do worry that I am going to be too old to meet someone and have a family when I finally get something permanent though.

The WISC community is even valued by the old guard. You know, the old white men who are the most senior and who everyone thinks can't change or don't want to. My old PI actually recommended me to get in touch with WISC. He said "Mira, I know I didn't always make it as easy for you as I should have done, and I definitely didn't for the ones who came before you but I know I need to do better. I need to learn more. I saw this and it looks like something worth being part of. Maybe they can help you where I couldn't."

My mentoring group are really close now. As well as the formal sessions with our mentor, we have a messaging group where we chat about everything else as well. We've helped each other by looking at fellowship applications and papers, and talking through major life decisions. One of us went for a major fellowship recently and seeing her do it makes me think that this is something that I can do soon. I don't think I would have even known about fellowships before, let alone how to write an application or what makes a good one, or know that I have a network of people who will check it over, give me feedback, and a mentor who is really senior in the field who can write me a reference.

I've actually offered to mentor my own group now and I know more of us mentees are doing the same – mine are all PhD students. It's an international group, and although sometimes the contexts they're in can be different, we all have that connection of being in the same field. It feels good to give back to WISC and to the community, and hopefully I can act as a role model to them, help them not to feel so alone, and inspire them like WISC and my mentor have inspired me.

SIX

Stories from STEM

In this chapter we share some of the stories that have been told within the autoethnography group, bringing you (the reader) these rarely heard voices through a focus on embodiment. The images and words were shared in meetings, emails, or instant messages. This chapter does not discuss these stories with references or citations to wider literature, as we have done in other chapters. Instead, we present them as snapshots of the lives of the women who took part in these studies and share them with each individual's consent.

Our intentions for the work were for it to reflect the experience of those engaged with the projects, and to find points that resonated with each other, wider members of WISC, and the scientific community. We wanted to find the human points of connection that allowed others to realise the impact and lived experiences we shared. In order to achieve this, we used a variety of methods (see Chapter Two for a discussion of methods used and why), with an emphasis on creativity and embodied experiences. All the chapters of this book bar Chapter One include fictional *vignettes* drawn from our data, and images from our research, and this one is no different. It was our choice to use fictional narrative to create *vignettes* that resonated with experiences in order to protect the anonymity of our participants. The *vignettes* draw on experiences shared within the survey, the reflective research group meetings, and the collaborative autoethnography. The team worked together to choose themes that resonated for us and we then disseminated these within the wider community to get feedback and input. In presenting these stories, we have chosen to take a chronological approach, sharing the themes as they arose naturally in meetings and discussion over the period of September 2020 to July 2021. Before the meetings started formally, many of the group discussed the broad topics that they thought might be relevant to the project, including the gendered nature of 'lab-safe clothing' and how it removes femininity; and the stupid things that people say to keep women out of labs.

Over time the group grew. From the initial plan to have 6 people sharing experiences, the ongoing project now includes 12 academics from the US, UK, and Europe. In addition to the stories shared here, they also discussed practices for managing their groups, tools for prioritising work, the idea and enaction of equity within their group, and how they could work to increase diversity and representation. In a paper detailing findings from the collaborative autoethnography triangulated with data from the WISC second survey and work with research groups, we wrote:

> The CA [collaborative autoethnography] group found the space to reflect, process, and share with a community of supportive peers. The meetings, rather than being another burden on their time, became points of connection and support. The importance of community for women in science is widely recognised.[1,2] Given the pre-COVID-19 context of the lack of diversity within chemistry,[3–5] it is little surprise that women, as they are often the main care givers within the home, have been impacted more by the coronavirus pandemic.[6,7] … The importance of community (or lack thereof through lockdown and social

isolation) was another important factor contributing to personal experiences. Communities and networks are vital for those who are marginalised.[2,8] Previous work with the supramolecular community demonstrated the value of reflexive and creative approaches to help build communities and networks,[9] allowing members to identify and disseminate their experiences to better understand the impact of marginalisation. The ongoing work with the CA and research groups supporting their reflective and reflexive processes[10] has been valued by all participants for the opportunity to share, connect, and feel less alone.[11]

COVID-19

Our original intention was to capture and explore the lived experiences of running a research group as a woman. As we wrote at the start of the project, we wanted to 'collectively craft the rhythms of our work and lives'. However, the timing of the project meant that, in addition to this, we also had the scope to capture and explore the lived experiences of being a female supramolecular chemist in academia through the 'unprecedented' COVID-19 pandemic (see Figure 6.1). At the time of writing the first draft of this book (20 July 2021), 67.5% of the UK population had had two doses of the COVID-19 vaccine (87.9% had had one dose),[12] and yet we were on the crest of a third wave of infections.[12] Globally, cases continue to rise, with variants of COVID-19 causing concern, particularly for those with underlying health issues.[13–15]

The group shared how the COVID-19 pandemic had affected them. There was a lot of fear about the unknown, with cases skyrocketing in many areas of the world. Most had been inundated with teaching and meetings, which had pivoted to being online, experiences that took longer and required more energy than in person (see Figure 6.2). One said she felt "like I'm in an episode of Dr Who as the laptop sucks my living

Figure 6.1: 'Unprecedented' sudden emergency

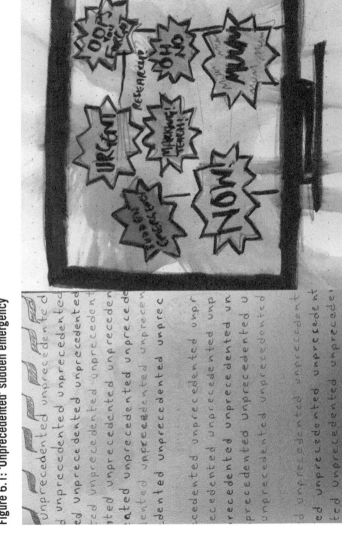

Figure 6.2: Being online

energy and my soul". Those who were doing face-to-face teaching talked about rules over wearing masks/visors, and we realised that regulations differed between different institutions, even those within the same country. Many had had annual performance/progression/review meetings cancelled and had been advised to prioritise teaching over all other academic work, but then had been asked why they were not submitting grant applications. They felt that the pressure was worse than usual, particularly as they were driven to ensure that the student experience was the same. Students were more disgruntled, and people shared how difficult it was to integrate new postgraduate students into a group when labs were split, and people were isolating. Others felt guilty for not teaching due to research time buyout (where their time has been externally paid for work on a particular project), or that they were in a period of calm before the storm of their teaching was to begin.

Some participants of this study had recently begun new jobs and shared how it was to be new in a 'weird time'. Everyone felt flat:

> COVID is feeling nearer to home and yet the world and people seem to be taking it less seriously. I am very distracted and anxious.
> Lacking in motivation and finding it hard to build a head of steam. I feel I get stuck and repeat a thought (or question) like a broken record rather than hearing the answer and moving on like I should.
> What do I want out the session today?
> Connection.
> To find a way to move forward.
> M o t i o n.
> I feel soft and weak when I want to feel hard and strong.

The pressures of work, and the needs of keeping the family together were at times overwhelming. One shared how she felt like a small dot between two giant rocks (see Figure 6.3).

Figure 6.3: I'm a small dot between two giant rocks

This feeling was echoed, particularly among those with young families. There was a lot of resonance with feelings of having to carry others and support people, while not being that good at asking for support in return. Resilience levels were low, and individuals felt as though there was no time for research.

Overwork

Everyone in the group felt the pressures of overwork. Although overwork is common within academia (see Chapter Three), these feelings were exacerbated through COVID-19. Everyone agreed that there was more pressure and stress than normal. COVID-19 was making things worse, and online schedules meant that people had less freedom. There was a level of mental gymnastics involved with COVID-19, which people had to deal with without the mechanisms in place that would normally offer support – whether that was interacting with other people or officially through the institutions.

> 'Universities only care about the money coming in not about the pressure it puts on you.'

The group shared feelings of guilt in every direction, and experienced pressure when they 'should' have been happy

> 'I landed a grant and I only feel pressure. I'm frozen with pressure.'

They shared that being a PI (Principal Investigator) was very stressful. Bottlenecks included applying for tenure in the US, where COVID-19 extensions were available, but it was not clear if extra time would, in turn, increase the level of requirements that needed to be met.

The group decided to focus on ways to prioritise and find balance. This was initiated after a meeting in which everyone shared experiences of having to continuously firefight and only being able to focus on what needed to be done immediately rather than being able to plan for the bigger picture. There was agreement that if people carried on the way they were working they would burn out. One person shared a memory of Mr Creosote from Monty Python, and how it reminded her of the line 'one wafer thin mint?', and instead today it was 'one more email' and then she would go.

KABOOM!!

The group shared how they would say yes to anything to get things done and for a quiet life. However, there was a lot of anger and frustration, and this was impacting on people's health. Coping strategies included lists, and lists of lists, but these were getting out of hand. One reflected on whether she was taking on work when she did not have to, questioning whether her tendency to say yes and take on more tasks and more responsibility was a distraction from getting on with the difficult stuff. Another, when asking her line manager for help and support, was told "you just have to prioritise".

There was a disconnect between the reality of living and working through COVID-19, and the ways in which they were able to live and work, and most importantly take care of themselves to sustain pre-pandemic levels of working.

Self-care, community, and celebrating small wins

As a group they decided to see whether focusing on each other's successes and sharing ways in which they looked after themselves would help them to focus on the positives. A beneficial change between lockdown 1.0 and 3.0 was an ability to delegate more, to be able to ask for help, although this was still hard for many, see Figure 6.4.

The group decided to meet and each bring an example of how they had been a 'badass' in the previous month, to share at least one nice thing they had done for themselves, and a silver lining from COVID-19 they had not been expecting. They found it easier than usual to focus on the positive.

Examples of badassery included:

'Standing up for myself and going as PI on an institutional grant.'

'Realising I have the capacity to keep going. ... But that I shouldn't push myself beyond where I can cope.'

'Saying no when work asked me to do something untenable.'

'Got invited to interview.'

'Stepping into big girl shoes on a big grant proposal.'

'Convinced boss to change product and ordered four boxes.'

'Published my 100th paper.'

'Won two grants.'

Examples of doing something nice included a lot of exercise, such as yoga, running, trampolining, walking, as well as making time,

Figure 6.4: Why is it so hard to put myself first?

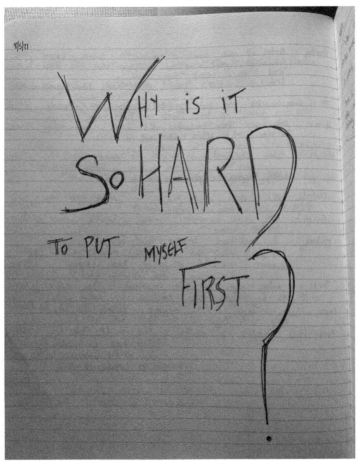

for example to have a bath, take leave, or have date nights. There is a wider discussion to be had that the basics of taking a bath or having time to exercise are relegated to 'self-care', or doing something nice, when they are relatively basic requirements.[16]

The final task of sharing a silver lining from COVID-19 was more varied. For those with children, having extra time

Figure 6.5: The benefits of working at home

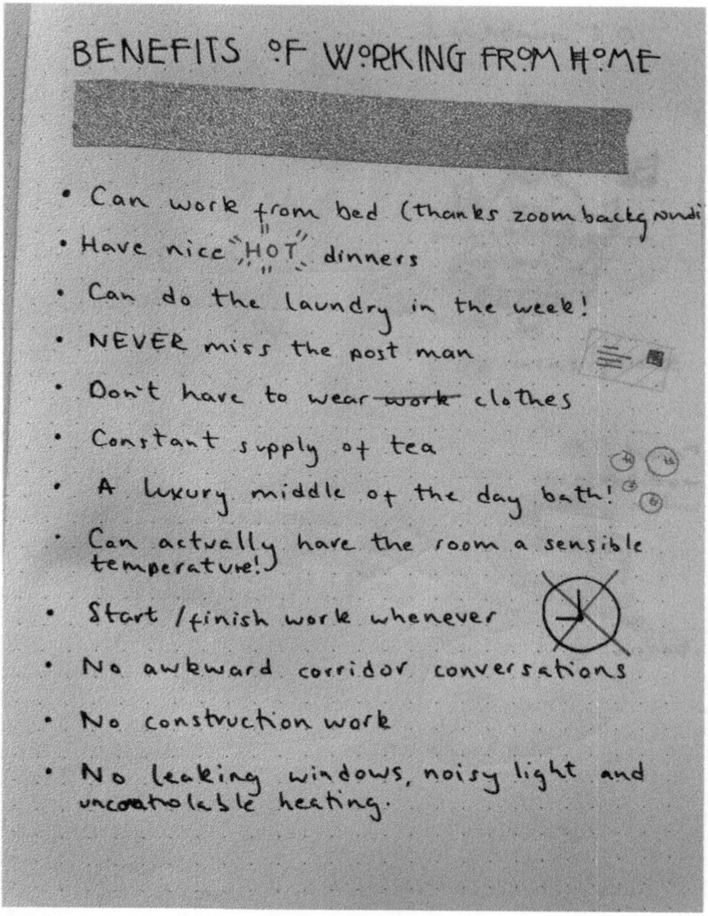

> **BENEFITS OF WORKING FROM HOME**
>
> - Can work from bed (thanks zoom backgrounds)
> - Have nice "HOT" dinners
> - Can do the laundry in the week!
> - NEVER miss the post man
> - Don't have to wear ~~work~~ clothes
> - Constant supply of tea
> - A luxury middle of the day bath!
> - Can actually have the room a sensible temperature!
> - Start/finish work whenever
> - No awkward corridor conversations
> - No construction work
> - No leaking windows, noisy light and uncontrollable heating.

with them was seen as a silver lining, although not without its own complications or challenges. Working from home was a positive (see Figure 6.5).

The need for community and connection was a big factor in the group's experience of COVID-19 and the project, and many other silver linings included an emphasis on communication, and contacting and connecting with others over the internet:

> 'Being able to keep in touch with people around the world.'

> 'Getting to know you all. Getting to know more people.'

> 'Networking to connect with people, especially over the internet.'

> 'People are more connected. I can speak to friends and lots of different groups of friends. I feel more connected with colleagues even though we don't see each other.'

Through the project, the group became close, maybe closer than expected, as they shared honest and authentic reflections of their lives. This connection allowed them to reveal things that they would not feel comfortable admitting to others in a work scenario (see Figure 6.6), especially around the idea of being successful when others were not.

Femininity and being a woman in the lab

The group were not always serious and, along with connection around the more sombre topics, they also discussed more light-hearted issues. We would of course not admit in print to conducting a statistically significant and totally unbiased comparative study of different luxury toilet paper brands while in global lockdown, but appropriate clothing that was worn in the laboratory was definitely talked about.

Conventionally in a chemistry laboratory it is imperative for health and safety reasons to ensure that along with wearing a lab coat, gloves, and safety goggles, all skin is covered up. This is for safety and practicality. On the feet this would mean, for example, no sandals or any gap above socks that exposes the ankles. This is in case there is a spillage of a caustic chemical, which would

Figure 6.6: This is how I feel sometimes, one flower having to stand tall while everything else is dying around me

lead to an injury if the chemical were to contact the skin. In addition, many labs mandate that students and staff wear natural fibres such as cotton, as these are less flammable. In practice, 'lab-safe' clothes can equate to being told to wear a typically masculine uniform of baggy cotton jeans, trainers, and the like. The argument for this is that clothes in the lab need to be able to be removed easily in case of spills. Adding into the equation oversized lab coats and hair that has to be tied back can mean that for young women being in the lab is somewhat equivalent to removing their femininity, if they have been used to expressing that femininity through their clothes. Although some may say that a lab is not a place to express personality and identity, this perspective homogenises lab workers, and expects them to all tend towards the 'average' – which of course is skewed by imbalanced representation. We all have personality and identity, but only some are told that this should not be expressed, or that it is the wrong sort of personality. In the labs where we have been, typical 'blokey' behaviour raised very few eyebrows, but overtly feminine behaviour had a different impact depending on the group.

The goal of equality is not to remove femininity in order for women to be considered on a par with men. This does not put people on a level playing field. Instead, it reinforces the idea that only some bodies (male) are allowed to exist within the space. Equality and inclusivity mean that the space needs to genuinely reflect diversity and be comfortable to all people who might inhabit it.

When Jen L first began spending time in a lab for a WISC research project, she agonised over 'what to wear', because prior to that she had delineated a 'work wardrobe' from her 'comfy at home wardrobe'. She asked the group what she should wear (which seemed an easier question than to think about the logistics and actuality of doing the research). This started a discussion that lasted through collaborative group meetings as well as less formal evening meet-ups. One member shared how, as a PhD student, she had had a stand-up argument with the health and safety officer in her lab as he bawled her out for wearing tights rather than the ubiquitous baggy jeans: "I told him that I could take my tights off quicker than he could pull his trousers down". A compromise was reached in that she was allowed to wear natural (that is, cotton/wool) tights underneath her lab coat.

This approach to lab-safe clothing was one adopted by other members of the group for their students. As more senior academics, the members themselves spent less time in the lab, and were thus less restricted in what they could wear. Some of their students shared with them how they attempted to retain their femininity – through wearing fancy earrings, or tying their hair up in elaborate styles.

The group also began collecting 'dumb things' they had been told about why they shouldn't pursue a career in chemistry, and many of these anecdotes such as "your hands are too small to hold a solvent bottle" or the classic "women can't stand at a fume hood because it will make their wombs sink" found their way into the fictional *vignettes* we have shared.

Emotional toll

The emotional impact of the pandemic was a frequent topic of conversation within the group. This was discussed from a personal perspective, as well as that of an academic and/or group leader:

> 'I am very distracted and anxious. Lacking in motivation and finding it hard to build a head of steam.'

> 'Someone died of COVID while lecturing in South America – is HE [Higher Education] worth dying for?'

> 'I have so much anger and I don't know why. I'm frustrated I don't have clean data. I can't sleep. I have so much anxiety.'

For those with young children, they were having to combine childcare with work. At more than one meeting young children were in attendance, and one member joked that she plugged her child into a tablet and stuffed her full of whatever snacks she asked for just to be able to work. These experiences were common for working mothers across the world, who felt a combination of exhaustion, panic, guilt, and anxiety.[17] School closures had an impact on parents' mental health.[18] The situation felt unsustainable:

> 'It's week four of homeschooling and the wheels are falling off. We were told today that schools won't open until early March at the earliest. I'm working 7am to midnight. ... I can't continue at this level, I am so rundown. On Friday instead of a nap I looked at paper drafts. I need a career-building momentum of papers after maternity leave. I was just getting momentum.'

These feelings were echoed in a poem by Cali Prince:[19(p37)]

Who was it
during
the COVID-19 pandemic

Who worked
day
until next dawn
at the hearth of the kitchen table?

The women,
they said.

Who was it
who home-schooled
their daughters,
their sons?

Who quelled
the disorder

and alleviated
anxieties

As they toiled
together?

Not in solitude,
not in silence
or peace
not in purpose-built studies
but in the rubble
of breakfast dishes
and half-written sentences?

Not always,
but most often
it was the women,
they said.

And in the midst
of the pandemic
whose productivity
was compromised,
whose attention was
divided and split?

Not always,
but most often
it was the women,
they said ...

The COVID-19 pandemic and lockdown had a negative impact on many people's mental health,[20] and academics were not immune from this.[21] Unsurprisingly, this was reflected in the participants of the collaborative autoethnography. In addition to their own struggles, they shared concerns about their groups' mental health:

'My group are so disconnected. We're doing Zoom meetings but the sessions are not so helpful.'

'I had a new international student start and I'm so worried that they aren't able to connect with anyone.'

'They're just fragmented and I don't know what to do to make it better.'

The group felt a high level of responsibility for their students, and this was an extra burden on top of their responsibilities for themselves, their families, their loved ones, and pressure to do work. The group discussed relationships inside and outside the lab. Managing groups and feeling responsible for their mental health and progress in addition to holding together for yourself and loved ones was a big theme, and one that we decided to explore with the WISC second survey (see Chapter Seven).

COVID-19 was tough on those in long-distance relationships if they were separated more than usual due to lockdowns and travel restrictions. For some it was a time when they actually got to be together, as they were able to work from home and live in the same place for once. "We lived together for two months – I felt bad for being so happy."

Fifth vignette Hermione, 38, Mid-career researcher

Working through COVID has been so bloody hard. The whole world has been struggling and I am no exception. However, in addition to being on this rollercoaster as an individual I have been working full-time throughout and been responsible for my whole group. I cannot tell you how many days I have had that have been full of meetings. Meetings with my department as we tried to sort out the sh*tstorm that was the 'pivot to online delivery'. That is not so easy in chemistry! I had it easier than some because I have always recorded my lectures, but I know that others had it so hard. It seemed to take four times as long to prepare for an online lecture. It's not just the planning, it's the recording, recording again, recording yet again, sorting out captions, sorting out the asynchronous material – the bits that the students have to read or watch alongside the actual lecture – then of course doing the synchronous bit. Have you ever sat in front of a screen and talked to little black boxes because everyone has their camera turned off? It's pretty soul-destroying and very hard on the voice! Some students actually contributed a lot more than if they'd been in a lecture theatre because they used the chat, but it is just so draining. And labs! I know of universities that sent out lab kits that students had to use at home – things like baking a cake or testing pH. To keep them engaged I guess. I am lucky in that we had actually done pretty much all our labs before lockdown and then got to plan how we would approach the next semester over the summer. Back-up plans and back-up plans and yet more back-up plans.

It takes time though. And I am also responsible for my group and making sure that we carry on getting the research out. I care about my group. I was really aware of not wanting them to feel abandoned and isolated so I made sure that we had regular meetings – in small groups as much as whole-group ones. They were a bit more informal too. I never used to wave hello at meetings before! I feel responsible for my groups' mental health. I am the one who has to hold them together and make sure they have work to do but aren't overloaded. I'm the one who has to negotiate for them to be in the lab again when they can. That was not fun. At first everything was just shut down for ages. Then other departments started to go back in and I had to negotiate for my lot to get in too. We were split for ages, so half were in at one time then the other half. It meant that my group weren't together as a whole for months. Communication started to break down, and I think some people got really isolated, particularly the newer ones. It was so hard to keep hold of everything. It's a bit better now, though our lab is on a schedule of two days on two days off. I've noticed that they're pushing themselves really hard – working really long days to get as much done as possible – and it has meant that we just can't do some of the reactions that take a long time.

They all had to move the fume hoods around and that was a palaver. My group used to be a lot smaller than the guy I share a lab with, but his has shrunk while mine has grown. That's what happens – it depends on grants and getting money in. He won't give me any more space though, and makes me feel like a demanding woman just for asking! Getting enough space is always a fight. One more thing that takes energy from me. I can't remember the last time I took leave. I feel like I need to be there all the time in case something goes wrong. That and the fact that it's not like I can go anywhere!

SEVEN

For the future

In this final chapter, we look to the future, and discuss the impacts of participating in this work for those involved. To date, everyone who has taken an active role in WISC has seen an increase in research outputs, grant successes, or career progression that is enabling them to achieve greater things within the field of supramolecular chemistry and beyond.

WISC's work is on-going. The collaborative autoethnography began life as part of two funded research projects – one that is exploring how women PIs can become better leaders, and another that is explicitly exploring how creative and reflective approaches can be used within groups to improve communication and result in an increase in both the amount and quality of the science produced. The second project is collecting data on groups' scientific outputs so that if there is a relationship between the approaches and scientific research, we will be able to demonstrate this quantitatively, as well as by using qualitative evidence of lived experiences. To date, we are creating a body of work that is already demonstrating the impact of community building and innovative interdisciplinary research between science and social science, and how these can change practice and policy within an academic disciplinary community. WISC's work is gathering pace. Since the funding of the two initial research projects described previously, we have received further funding to focus on expanding the model and framework of WISC into Africa, with the aim of increasing the visibility of Black women in science, and to support those who wish to further develop WISC as a part of independent fellowship awards. In addition, despite having been in existence formally for only a year and a half, WISC was put forward for three EDI prizes in 2021.

Triangulating data

Findings from WISC's second survey, which set out to explore the lived experiences of supramolecular chemists through COVID-19, have been triangulated with data from working with research groups in reflective meetings and data obtained through the collaborative autoethnography sessions. These results have been published in *CHEM*.[1],* The survey, which had 105 respondents from 6 continents, found that while people across all career stages were impacted by COVID-19, those at the later stages with responsibility for groups, particularly those who had caring responsibilities, were most likely to be affected negatively. Conversely, the groups that reported COVID-19 being a positive and/or productive time were most likely to be PhD students or post-docs without caring responsibilities. The paper shared the emotional toll on those who were responsible for leading and managing groups through periods of lockdown and then returning to labs.

The return to labs was interesting, in that different institutions handled the social distancing requirements very differently. Some asked group leaders to split their researchers, with only half the group accessing the lab at any one time, but allowing the group as a whole to have continuous access. Others put

---

* Some of the data in this chapter has been previously published in *CHEM*,[1] and is reproduced with kind permission from Cell Press and Elsevier.

whole groups onto a rota with others, so that the group only had access to the lab half the time, but the researchers within that group were all together for the time they had. Many students shared that returning to the labs was hard:

> It has been immensely emotionally draining in returning. It requires additional effort each day to focus on work while the numbers are/were rising so steadily. (PhD student, woman)

> Finding return to labs hard, difficult to use the shared equipment safely without feeling stressed. I feel pressure to make up for lost time so the return to labs has been very busy. (PhD student, woman)

> Our supervisor chose who got to go back and only informed those individuals at the last minute, leaving everyone else wondering if they are to stay at home or just haven't received his email yet. Post-docs got to go back first and didn't take the time to talk to the rest of the group. (PhD student, woman)

Our survey found that researchers and group leaders with split groups had a much harder time. Groups became more dissociated. One collaborative autoethnography member said "my group are divided", and many of the group sessions were spent talking about ways in which to increase group unity and communication when they were not able to be together in the same space.

The worry of carrying responsibility for a whole group through the pandemic was challenging for everyone. Other group leaders shared similar experiences, again particularly those who had additional caring responsibilities at home:

> We were increasing told to look after the students' mental health and make sure they were ok above everything. No one asked how the staff were coping. ... Returning

to labs has made dealing with PhD students a bit better, less reliant on me to tell them what to do. But they continually pester me asking when the rota will change (we have split the group in half and are doing week in and then week out). This isn't helped when other labs are in 100% of the time due to them having larger labs or smaller groups. Head of school says there is no more space, seems unlikely, just poorly organised space. ... We have been told to prioritise teaching from up high, but also from college we are asking where all the grants are, where are the papers since we've had all this extra time. (Independent researcher, woman)

I feel that my research team and I responded to the challenges that COVID posed to us with resilience and agility, the main impact to myself was that an amount of my personal and emotional resource was needed to support others and ensure the productivity of my team was maintained. This has left me drained and exhausted. (Independent researcher, woman)

I worked from home. Its arguably the hardest thing I've ever done. I have two kids (age 8 and 12) and they are somewhat autonomous, but it was still difficult. I worried about everything ... especially the well-being of my group, and of course our productivity which effectively fell to zero. (Independent researcher, man)

I have a leadership role in the Dept [department] and trying to sort out how to handle COVID in the best way for the Dept, students, and university has required a huge amount of time, effort, and anxiety, unrelentingly, for more than 12 months. The most negative effects for my research, besides the huge amount of extra work for me that COVID has necessitated to sort out Dept & university matters, is not seeing individual members of my group in person, other than by Zoom, for more than a year. Without face-to-face interaction one cannot really understand how everything is going, personally and professionally, for everyone and on particular projects. (Independent researcher, man)

These quotes demonstrate that our collaborative auto-ethnography group was not having isolated experiences within the field of supramolecular chemistry, and it is our belief that these same sentiments are likely to resonate through many other lab-based disciplines.

We concluded the article by writing about the process of conducting the research and participating in the collaborative autoethnography and/or reflective group sessions:

The challenges that students, post-docs, and independent researchers faced in supramolecular chemistry are likely to echo those faced by academics and researchers across not only the many fields of chemistry, but other disciplines too. What is novel in our approach is that findings presented herein are data triangulated from three sources, together with the use of a community specific group to address these challenges. As such, rather than looking at the problems from the outside, we as a community are exploring these issues as a means to address them. There was a negative impact of rotas on the mental health, communication, and productivity of research groups. Having caring responsibilities was the largest factor for all participants regardless of age, career stage or gender. The emotional load of managing a research group through COVID-19 was an unexpected burden borne unevenly across the academic community, falling as it does predominantly on those in STEM who are more senior in their careers, and who are thus more likely to also shoulder additional senior management responsibilities. This last factor, largely unrecognised by universities, without doubt contributed to the decision made by some women survey respondents to leave academia during/as a result of the COVID-19 pandemic.

> 'I have a nursery age child and not childcare or any family nearby so I basically couldn't do my job, which was increasingly more difficult with managing PhD students who couldn't go into lab ... I told my line manager about my lack of ability to do my job, and he just told me to make a note of it for our PDRs, which have now been cancelled.' (Independent researcher, woman)

> 'Upon returning to lab, I had lost all motivation to work. This event contributed quite strongly to my decision to leave academia.' (Other, woman)

If chemistry and science are to continue to tackle the EDI crisis, then it is imperative that the impact of COVID-19, particularly on those with caring responsibilities, lab groups, and who are from minorities where progression is limited, is ameliorated.

It will be necessary to track whether the long-term impact of COVID-19 increases the attrition of women from chemistry and decreases the progression of minority groups.[2-4] We suggest that a major tool in the arsenal used to address the lack of gender balance and diversity in science generally, and the impact of COVID-19 on those groups specifically, is establishing and growing networks of area-specific communities. This allows space for individuals to reflect on and share their lived experiences so that they are less isolated and marginalised. We offer WISC as a blueprint or model of how this may be achieved in an area-specific field that can be replicated across disciplines, borders, and communities.[1]

Using community to change culture

WISC aspire to use community, and the building of communities, in order to change culture, to inspire, and to reach out to others. The work we have developed has allowed us to build 'solidarity, friendship, and the development of supportive networks both in "real life" and online [which have played] an important role in feelings of connectedness, academic community making, and can also be beneficial to women navigating the gendered community'.[5(p95)] Our approach of regular meetings giving members opportunity to share and reflect is mirrored in the 'Group' described by Ellen Daniell in *Every Other Thursday*.[6] We, like Daniell's 'Group', aim to create an alternative to the old boys' club version of academia we shared in Chapter Three:

> a commonly used informal description of the upper echelons of academia is a network of 'old boys clubs' ... these 'old boys clubs' remain highly effective for their members, exclusionary to women, and ... play a tacit role in recruitment and selection, and the furtherance of [some] men's academic careers to the detriment of their women counterparts.[5(p91)]

By doing so, we aim to change the culture in chemistry and STEM which states that 'women leaders are not common'.[7(p62)] Instead, we aim to build a community that looks after its own, and that supports women and other marginalised genders and groups.

We are not passive in this work, as demonstrated by WISC's targeted clusters, each looking to provide support that is intersectional to those who face additional barriers due to the caring responsibilities of being a parent, having a disability, chronic illness, or neurodivergence, or being the first generation into higher education. Our clusters are open to everyone, regardless of gender or any other characteristic. They are there so that people can learn how to support themselves, and also how to support others, and to make informed choices from a position of knowledge. The choice of whether to have children or not is an example of this. In Chapter Three we discussed the fact that so few women in supramolecular chemistry (or science) reach the most senior career levels and also have a family. While all women to some degree experience 'the guilt of not doing the right thing, of making the right decision, [which] weighs heavily on our shoulders. If we decide to be mothers, we face the motherhood penalty',[8(p130)] women who want to pursue a career in science have very few role models or people who have carved out a path for them to follow.

We are thankfully moving towards a more inclusive and diverse community in chemistry. Professional bodies like the RSC and funders such as Wellcome and the Royal Society have all stated explicitly their desire to increase diversity across the sciences. We have come a long way from the early 1920s, when scientific papers declared the toxicity of women at certain times of the month[9] and used this as a reason to keep them away from science and other arenas: 'in the 1920s Dr Béla Schick believed that menstruating women produced a toxin called menotoxin which could wilt otherwise normally thriving flowers'.[8(p41)] Although this theory is fortunately no longer accepted, there is a long history of women being vilified: 'the female body has been imagined and put forward as a site of madness and badness since the time of Hippocrates'.[8(p14)] We still see similar arguments put forward to suppress and discriminate against those who are marginalised, for example due to being trans, brown, or Black.

When it comes to including those who are disabled, chronically ill, or neurodivergent in the lab, things are still murky. Laboratory health and safety regulations mean that there are certain restrictions on what can happen in the lab. As we saw in Chapter Six, this can impinge on freedom of choice when it comes to clothing. In addition, synthetic chemistry demands that a researcher has use of their hands, legs, and adequate vision to do the job and be aware of their surroundings. Customs and practices demand that fumehoods and benches are fixed at heights suitable for the 'average' researcher. In other words, they are designed for the average man. This can mean that labs are less accessible to those with disabilities, chronic illnesses, and neurodivergences, or for anyone who is smaller or taller than the average. They do not, as standard practice, have facilities that are wheelchair accessible, or that even contain comfortable seats that people can sit down and work on. Instead, hard wooden stools shaped to the average man's buttocks are the norm. Labs are pressured working environments that are often noisy, and have little privacy. WISC's Disability/Chronic Illness/Neurodivergence Cluster has recently begun exploring a project of work that is looking to design a virtual accessible and inclusive lab of the future funded by the RSC. Through consulting with chemists who currently work in labs, and who have a disability, chronic illness, or neurodivergence, the cluster will create a 3D virtual lab that can be used to raise awareness of the issues that face people in the current standard set-up. This project emerged from a discussion after a member thought that she had broken her leg, and wanted to know whether she would be allowed into the lab with a cast or in a wheelchair (the answer was no, due to health and safety concerns). The future lab would potentially include accessible seating, fumehoods that span floor to ceiling with adjustable benches to accommodate those of different heights, comfortable stools, sound-proofed areas for writing up, while remaining close enough to visually observe experiments, and platform-sharing technology so that automated experiments could be run remotely with no requirement to be physically present just to press a button to start a machine.

It is likely that there will be push-back against changing labs to make them more accessible and inclusive. There will be cost and resource implications, and it would go against the way things have always been done. Although universities have a duty in the UK to put reasonable adjustments in place, accessibility has not often had to be a consideration within labs because of the lack of people requiring (or disclosing that they require) such adjustments. If no one is asking for changes to be made, then that can be taken to mean that there is no demand for those changes. This conceals the needs of those who do not fit to the norm. The lack of diversity creates a culture which either does not welcome inclusion or fails to recognise where it does not welcome inclusion. This is a vicious cycle that results in systemic discrimination against those who are different.[10–12] There is also the question 'but where do we draw the line?' Legitimately, some careers are just not suitable for some people. If you are blind, can you be a driving instructor? If you have severe physical disabilities, is pursuing a career in a synthetic chemistry lab the most realistic option, or could you instead focus on computational aspects of chemistry? The challenge is ensuring that such decisions are made by the individual in their own best interests, and not forced on them by an environment that is inaccessible to their reasonable needs and that health and safety is not weaponised to keep them out. Are those with disabilities, chronic illnesses, and neurodivergences not in the lab because they do not see it as a place they are welcome? Or are they there, and just not disclosing, and so putting up with uncomfortable working conditions?[13]

The fact that labs were designed for the average man would come as no surprise to Caroline Criado Perez, author of *Invisible Women: Exposing Data in a World Designed for Men*. At the end of her book she writes:

> The solution to the sex and gender gap is clear: we have to close the female representation gap. When women are involved in decision-making, in research, in knowledge production, women do not get forgotten. Female lives and perspectives are brought out of the shadows. This is to the benefit of women everywhere..[and] is often to the benefit of humanity as a whole.[14(p318)]

WISC's work to support the retention and progression of women will help them to be involved in research and knowledge production, and to become senior enough in their careers to be involved in decision making. However, being involved in WISC will not be a guarantee to success. The current climate in higher education and research is constrained due to the COVID-19 pandemic and wider financial context, and as such academia, along with many other industries, has become even more intensified in its competitiveness. The numbers of students graduating with a PhD is far in excess of opportunities in academia, whether they are postdoctoral or permanent positions. Post-docs in particular can be very precarious,[15,16] and experiences vary depending on the individual PI[17] or training opportunities.[18] Critiques of the system and options for post-docs have been suggested.[19–21] Mentoring programmes such as those offered through WISC, along with the support clusters and community building, can work to ameliorate the isolation and anxiety faced by many at this career stage, which could lead them towards leaving academia and/or science completely.[2]

Collectively crafting the rhythms of our lives

Throughout this book, and in all the work that WISC is involved in, we have consciously done our best, in the words of A. Lin Goodwin, to:

> support other women, engage feminist practices that consciously centre women in the story – for example, make an effort to cite women scholars or tap women for ideas; assess your participation structures to ensure that women get the floor as often as men; and mentor women coming behind you, forward them for opportunities to lead and grow ... Notice inequity and challenge the normative. Learn to unsee the taken for granted.[22(p82)]

By employing feminist research practices[23] and utilising Embodied Inquiry[24] we have shown how breaking out of disciplinary norms to take an embodied and authentic approach to research allows the heretofore invisible and hidden experiences to become stories that touch and evoke responses. It has allowed us to reflect on research practices, to process, to listen, and to enact change. We have been able to pay attention to the rhythms of our work and our lives, and to take time to notice whether we want our individual situations to change, and how. It has been a collaborative journey, not a journey we undertook individually. In the words of the African proverb: 'if you want to go fast, go alone. If you want to go far, go with others'. In WISC Board meetings we have often reflected that it is the combination of people and skills that has allowed us to achieve so much in so little time, and on top of our day-to-day academic and personal commitments. We devote time to this work because we are passionate about it, and proud of it. We are proud to say that we are intersectional feminists,[25] and proud to say that we will fight for women and other marginalised genders or groups.

> being a feminist means advocating for gender equality and equity, fighting for justice for women and girls, actively intervening in and speaking out against unjust practices that demean women and hold them back, and being conscious of male privilege and leveraging that toward equity and fairness for women and girls. If everyone were a feminist, we could possibly make sexism, misogyny, and gender bias our history, not our continuous present.[22(p82)]

According to the Cambridge Dictionary, crafting means 'the activity or hobby of making decorative objects with your hands; or the activity of skilfully creating something such as a story'.[26] Together we have crafted WISC, the work of WISC, and the outputs of WISC. In doing so we have shared with each other the details and intricacies of our lives inside and outside the lab. We discovered where our experiences echoed and resonated with each other, regardless of age, disability, religion, ethnicity, or where in the world we lived. Pragya Argawal wrote: 'sometimes lives only take meaning when we look at them from the outside. When we are inside them, they often seem just ordinary'.[8(p2)] Our lives are ordinary to us. Throughout this book we have shared aspects from our lives in the hope that these ordinary experiences of discrimination due to gender can be more widely recognised and eliminated. Using our stories together in order to craft other stories, such as the wider narrative here and the individual fictional *vignettes*, will hopefully allow others to recognise themselves and their own experiences, as well as helping those who have not lived with such marginalisations to learn about the barriers, challenges, and joy of being a woman in supramolecular chemistry.

In Chapter One we set out the reasons why we wanted to write this book. In Chapter Two we shared the details of our methodological approach, and how we created the content and research data that has informed us. In Chapter Three we discussed the challenges of building an academic identity as a woman in STEM. The picture of gender marginalisation we presented in Chapter Four can be depressing. In some ways there has been little progress or change for many years, with women today experiencing the same issues around work–life balance and discrimination faced by women in the 1970s.[27] The reasons Jen L left her PhD in chemistry in the early 1990s[28] are still reasons why women do not see chemistry as a sustainable or attainable career, so leave this career path in droves.[2] However, excitingly there is now a willingness to change within the community. To listen to and learn from lived experience rather than deny it. To speak out rather than allow discrimination of others. This is evidenced by the publication of our EDI work in leading international chemistry journals and magazines.[29,30] Chapter Five related the story of WISC, and the work we are engaged in, and Chapter Six shared stories from the ongoing collaborative autoethnography project which ran throughout the COVID-19 pandemic. In this chapter we have shared insight and data from WISC's second survey and set out how we believe that building a sense of kinship and community is changing the supramolecular chemistry community for the better.

On a larger scale, we believe that our experiences in the field of supramolecular chemistry, the framework we have developed, and the lessons we have learned, can be used and adapted for other chemistry and STEM disciplines where there is a gender imbalance or marginalisation. Our hope is that those who read this book, whether they are researchers, research leaders, science administrators, funders, university or industry leaders, will listen to and learn from our lived experience, and work together to craft inclusive change. If we were asked what we would want for the future, we would want a scientific community where everyone is given the opportunity to learn and to progress, regardless of their gender, religion, ethnicity, race, sexuality, disability, or any other protected characteristic. People would be free to choose whether they wanted to have children or not with no detriment to their careers, and could undertake work they were passionate about in environments free from harassment, in locations that suited their home and/or family life without pressure to relocate or negotiate the 'two body problem'. It would be a scientific community that valued the contributions and wellbeing of its workers equally, and where people worked in collaboration rather than competition to produce knowledge that could make the world a better place.

Sixth vignette Phyllis, 63, Senior researcher

It was different in my day. There just weren't enough women about to have any sense of a community. I can see the difference it makes for the younger ones coming through to have enough of a critical mass to make a difference to each other. It felt very lonely back then – still does in a way. I think choices were starker – you either had to decide career *over* family, or career *after* family although that could mean that you just never progressed as far as your

FOR THE FUTURE

male colleagues. I don't have children. I've had relationships, but never met a man that could cope with me prioritising my career in the way I had to. There have been so many women and men doing work to change things. It does make me wonder *why* young women are still facing the same barriers and having to make the same choices we did 30 years ago, but I suppose at least there are more of them making them. I do think there's a real willingness to learn and to do things differently though. I mean, I can count the number of times I *haven't* heard some version of 'you only got that because you're a woman' when I've shared funding or publication success, or when I was made a Fellow. I think the more of us who are senior can stand up and shout and lend our names and support to the ones coming through the better. I mentor where I can – both officially and unofficially. What makes me so proud is that I see my male colleagues doing the same. Really championing young women. I know we need to address all kinds of diversity in chemistry – goodness knows we don't have enough range of skin colour – but I am hopeful. I appreciate that they recognise the hard work we all put in – by doing our best to change things and in some ways just by being here and achieving what we have achieved. At least they have role models of a sort – more than I ever did. I want to be hopeful. I see all these brilliant young women, and I see them *not* dropping out, not leaving to a different profession, but staying, and having families, and getting the grants, and getting the papers out. One showed me a mug the other day that her wife had bought her – what did it say now? Oh yes – 'Girls just want to have fun*ding for scientific research*!' I thought that was a hoot! I am sometimes amazed by how bold this new generation are, how brave for calling out behaviour they will not put up with. I'm not sure that I could have back then – it felt much too risky. A lot more was just accepted as well.

If you asked me what I want for this new generation? I would say I want them to keep on being brilliant, to keep on being bold, to have ideas and to challenge us old dinosaurs. I love the way they smile and wave and get us all on side. It makes us *want* to work with them, and it keeps us on our toes as well! I want them to have opportunities – for scientific research, to have families, to have a life, and to have fun. I want them to remember that academia won't love you as much as the people in your life will. That it won't necessarily reward their long hours or dedication. It will take and take and take. I've heard it described as a 'consensual abusive relationship' and that hit home a bit! I want those coming through to have healthy, meaningful relationships with their work and the rest of their lives. I want them to keep fighting for themselves and for each other and to really change the culture of chemistry and science so that no one is marginalised anymore, and so that what *really* matters is the science.

References

Chapter One

1. R. Colwell and S. Bertsch McGrayne, *A lab of one's own: One woman's personal journey through sexism in science*, Simon & Schuster, New York, 2020.
2. E. Daniell, *Every other Thursday: Stories and strategies from successful women scientists*, Yale University Press, New Haven, CT, 2006.
3. S. Rosser, *The science glass ceiling: Academic women scientists and the struggle to succeed*, Routledge, New York, 2004.
4. S. Rosser, *Breaking into the lab: Engineering progress for women in science*, New York University Press, New York, 2012.
5. S. Rosser, *Academic women in STEM faculty*, Palgrave Macmillan, Basingstoke, 2017.
6. V. Gornick, *Women in science: Then and now*, The Feminist Press, New York, NY, 25th Anniversary, 2009.
7. C. Caltagirone, E. Draper, C. Haynes, M. Hardie, J. Hiscock, K. Jolliffe, M. Kieffer, A. McConnell and J. Leigh, An area specific, international community-led approach to understanding and addressing EDI issues within supramolecular chemistry, *Angewandte Chemie International Edition*, 2021, 60, 11572–11579.
8. L. Mullings, *On our own terms: Race, class, and gender in the lives of African American women*, Routledge, New York, 1997.
9. E. Monosson, *Motherhood, the elephant in the laboratory: Women scientists speak out*, Cornell University Press, Ithaca, NY, 2008.
10. S. Ahmed, *Living a feminist life*, Duke University Press, Durham, NC, 2017.

REFERENCES

11. RSC, *Breaking the barriers: Women's retention and progression in the chemical sciences*, Royal Society of Chemistry, London, 2019.
12. A. Brecher, Balancing family and career demands with 20/20 hindsight, in *Motherhood, the elephant in the laboratory: Women scientists speak out*, ed E. Monosson, Cornell University Press, Ithaca, NY, 2008, pp 25–30.
13. E. Chuck, Alyssa Milano promotes hashtag that becomes anti-harassment rallying cry, *NBC News*, 2017, https://www.nbcnews.com/storyline/sexual-misconduct/metoo-hashtag-becomes-anti-sexual-harassment-assault-rallying-cry-n810986 (accessed 2 March 2022).
14. Everyones Invited, Everyone's Invited, www.everyonesinvited.uk (accessed 15 July 2021).
15. R. Younger, Everyone's Invited names almost 3,000 schools following claims of sexual assault, rape and harassment, *ITV News*, 9 June 2021, https://www.itv.com/news/2021-06-09/everyones-invited-names-almost-3000-schools-following-claims-of-sexual-assault-rape-and-harassment (accessed 2 March 2022).
16. S. Osbourne, Everyone's Invited: What is the website with 10,000 sexual abuse complaints and will there be an inquiry? 30 March 2021. https://www.independent.co.uk/news/education/education-news/everyones-invited-school-sexual-abuse-what-b1824332.html (accessed 2 March 2022).
17. S. Elliott, 30 schools in Edinburgh and Lothians named in Everyone's Invited sexual assault testimonies, *Edinburgh Evening News*, 12 July 2021, https://www.edinburghnews.scotsman.com/education/30-schools-in-edinburgh-and-lothians-named-in-everyones-invited-sexual-assault-testimonies-3303141 (accessed 2 March 2022).
18. Rape Crisis England & Wales: Statistics on sexual violence, https://rapecrisis.org.uk/get-informed/about-sexual-violence/statistics-sexual-violence/ (accessed 15 July 2021).
19. UK. Parliament, Serious fall in rape prosecutions examined with Victims' Commissioner for England and Wales, https://committees.parliament.uk/committee/83/home-affairs-committee/news/156062/serious-fall-in-rape-prosecutions-examined-with-victims-commissioner-for-england-and-wales/ (accessed 20 July 2021).

20. C. Barr and A. Topping, Fewer than one in 60 rape cases lead to charge in England and Wales, *The Guardian*, 23 May 2021, https://www.theguardian.com/society/2021/may/23/fewer-than-one-in-60-cases-lead-to-charge-in-england-and-wales (accessed 2 March 2022).
21. Cameron Kimble, Sexual assault remains dramatically underreported, Brennan Centre for Justice, https://www.brennancenter.org/our-work/analysis-opinion/sexual-assault-remains-dramatically-underreported (accessed 2 March 2022).
22. N. Brown and J. Leigh, *Ableism in academia: Theorising experiences of disabilities and chronic illnesses in higher education*, UCL Press, London, 2020.
23. S. Ahmed, *On being included: Racism and diversity in institutional life*, Duke University Press, Durham, NC, 2012.
24. M. Beard, *Women & power: A manifesto*, Profile Books, London, 2017.
25. Jenny Jones, The outrage at my suggestion of a 6pm curfew for men exposes a depressing reality about violence against women, *iNews*, 15 March 2021, https://inews.co.uk/opinion/the-outrage-at-my-suggestion-of-a-6pm-curfew-for-men-exposes-a-depressing-reality-about-violence-against-women-914567 (accessed 2 March 2022).
26. G. MacInes, I went to Sarah Everard's vigil to pay my respects to a woman who could have been any of us, *Global Citizen*, 19 March 2021, https://www.globalcitizen.org/en/content/sarah-everard-vigil-clapham-gender-violence-uk/ (accessed 2 March 2022).
27. RSC, *Diversity landscape of the chemical sciences*, Royal Society of Chemistry, London, 2018.
28. W. Joice and A. Tetlow, *Disability STEM data for students and academic staff in higher education 2007/08 to 2018/19*, Royal Society, London, 2021.
29. P. Fara, *A lab of one's own: Science and suffrage in the first world war*, Oxford University Press, Oxford, 2018.
30. United Nations: Sustainable development goals, www.un.org/sustainabledevelopment/sustainable-development-goals/ (accessed 15 July 2021).
31. C. Flaherty, Inside higher education, www.insidehighered.com/news/2020/04/21/early-journal-submission-data-suggest-covid-19-tanking-womens-research-productivity (accessed 10 November 2020).

References

32. B.P. Gabster, K. van Daalen, R. Dhatt and M. Barry, Challenges for the female academic during the COVID-19 pandemic, *Lancet (London, England)*, 2020, 395, 1968–1970.
33. V. Evans-Winters, *Black feminism in qualitative inquiry*, Routledge, Abingdon, 2019.
34. S. Ahmed, *Complaint!*, Duke University Press, Durham, NC, 2021.
35. WISC, The International Women in Supramolecular Chemistry Network.
36. J.S. Leigh, J.R. Hiscock, S. Koops, A.J. McConnell, C.J.E. Haynes, C. Caltagirone, M. Kieffer, E.R. Draper, A.G. Slater, K.M. Hutchins, D. Watkins, N. Busschaert, L.K.S. von Krbek, K.A. Jolliffe and M.J. Hardie, Managing research throughout COVID-19: Lived experiences of the supramolecular chemistry community managing their research through COVID-19, *CHEM*, 2022, 8(2), 299–311.
37. WISC, *WISC 1st Skills Workshop*, 2021, https://d017147e-00c2-4e9e-88fd-affbf75f13ce.filesusr.com/ugd/e3c05f_ef7a72f4eaeb4dc58f83e3c84dee4f36.pdf (accessed 2 March 2022).
38. J. Blanden, C. Crawford, L. Fumagalli and B. Rabe, *School closures and parents' mental health*, Institute for Social and Economic Research, University of Essex, 2021.
39. M. Barnes, You can't un-see colour, a PhD, a divorce, and the Wizard of Oz, in *Feminism and intersectionality in academia: Women's narratives and experiences in higher education*, ed S. Shelton, J. Flynn and T. Grosland, Palgrave Macmillan, Basingstoke, 2018, pp 155–166.
40. S. Stiver Lie and V. O'Leary, *Storming the tower: Women in the academic world*, Kogan Page, London, 1990.

Chapter Two

1. S. Rosser, *Feminism within the science and health care professions: Overcoming resistance*, Pergamon Press, Oxford, 1988.
2. S. Rosser, *Academic women in STEM faculty*, Palgrave Macmillan, Basingstoke, 2017.
3. M.A. Mason and E.M. Ekman, *Mothers on the fast track: How a new generation can balance family and careers*, Oxford University Press, Oxford, 2007.

4. V. Gornick, *Women in science: Then and now*, The Feminist Press, New York, 25th Anniversary, 2009.
5. N. Denzin, *The qualitative manifesto*, Routledge, Abingdon, 2010.
6. J.H.I. Stanfield, The possible restorative justice functions of qualitative research, *Interntation Journal of Qualitative Studies in Education*, 2006, 19, 673–684.
7. C. Caltagirone, E. Draper, C. Haynes, M. Hardie, J. Hiscock, K. Jolliffe, M. Kieffer, A. McConnell and J. Leigh, An area specific, international community-led approach to understanding and addressing EDI issues within supramolecular chemistry, *Angewandte Chemie International Edition*, 2021, 60, 11572–11579.
8. U. Flick, *The SAGE handbook of qualitative data collection*, Sage, London, 2018.
9. D. Silverman, *Interpreting qualitative data*, Sage Publications, London, 3rd edn, 2006.
10. H. Chang, F. Ngunjiri and K.-A. Hernandez, *Collaborative autoethnography*, Routledge, Abingdon, 2016.
11. R. Kennelly and C. McCormack, Creating more 'elbow room' for collaborative reflective practice in the competitive, performance culture of today's university, *Higher Education Research & Development*, 2015, 34, 942–956.
12. J. Leigh and N. Brown, *Embodied inquiry: Research methods*, Bloomsbury, London, 2021.
13. V. Evans-Winters, *Black feminism in qualitative inquiry*, Routledge, Abingdon, 2019.
14. S. Wilson, *Research is ceremony: Indigenous research methods*, Fernwood Publishing, Halifax, 2008.
15. H. Kara, *Research ethics in the real world: Euro-Western and indigenous perspectives*, Policy Press, Bristol, 2018.
16. A. Clark and P. Moss, *Listening to young children: The mosaic approach*, National Children's Bureau for the Joseph Rowntree Foundation, London, 2001.
17. A. Clark and P. Moss, *Spaces to play more listening to young children using the mosaic approach*, National Children's Bureau, London, 2005.
18. H. Kara, *Creative research methods in the social sciences: A practical guide*, Policy Press, Bristol, 2015.

REFERENCES

19. P. Leavy, *Method meets art: Arts-based research practice*, Guildford, New York, 2nd edn, 2015.
20. R. Coleman, *Glitterworlds: The future politics of a ubiquitous thing*, Goldsmiths Press, London, 2020.
21. N. Brown and J. Leigh, in *Theory and method in higher education research*, eds J. Huisman and M. Tight, Emerald, Bingley, 2018, vol 4.
22. V. Braun and V. Clarke, Using thematic analysis in psychology, *Qualitative Research in Psychology*, 2006, 3, 77–101.
23. M. MacLure, *Discourse in educational and social research*, Open University Press, Buckingham, 2003.
24. P. Leavy, *Fiction as research practice: Short stories, novellas, and novels*, Routledge, Abingdon, 2016.
25. P. Clough, *Narratives and fictions in educational research*, Open University Press, Maidenhead, 2002.
26. B. Bjørkelo, Workplace bullying after whistleblowing: future research and implications, *Journal of Managerial Psychology*, 2013, 28, 306–323.
27. R. Philips and H. Kara, *Creative writing for social research: A practical guide*, Bristol University Press, Bristol, 2021.
28. M. Hammersly, Ethnography: Problems and prospects, *Ethnography and Education*, 2006, 1, 3–14.
29. S. Pink, *Doing sensory ethnography*, Sage, London, 2009.
30. A. Bochner and C. Ellis, *Evocative autoethnography: Writing lives and telling stories*, Routledge, London, 2016.
31. A. Bleakley, From reflective practice to holistic reflexivity, *Studies in Higher Education*, 1999, 24, 315–330.
32. J. Leigh and R. Bailey, Reflection, reflective practice and embodied reflective practice, *Body, Movement and Dance in Psychotherapy*, 2013, 8(3), 160–171.
33. J. Leigh, *Conversations on embodiment across higher education: Teaching, practice and research*, Routledge, Abingdon, 2018.
34. C. Ellis, Fighting back or moving on: An autoethnographic response to critics, *International Review of Qualitative Research*, 2009, 2, 371–378.
35. I. Wellard, *Researching embodied sport: Exploring movement cultures*, Routledge, London, 2015.
36. R. Merton, *On social structure and science*, University of Chicago Press, Chicago, IL, 1996.

37. L. Dubois Baber, Color-blind liberalism in postsecondary STEM education, in *Diversifying STEM: Multidisciplinary perspectives on race and gender*, ed E. McGee and W. Robinson, Rutgers University Press, New Brunswick, NJ, 2020, pp 19–35.
38. S.A. Southerland, G.M. Sinatra and M.R. Matthews, Belief, knowledge, and science education, *Educational Psychology Review*, 2001, 13, 325–351.
39. B. Latour, *Pandora's hope: Essays on the realities of science studies*, Harvard University Press, Cambridge, MA, 1999.
40. K. Barad, *Meeting the universe halfway: Quantum physics and the entanglement of matter and meaning*, Duke University Press, Durham, NC, 2007.
41. T. Shin, Real-life examples of discriminating artificial intelligence, https://towardsdatascience.com/real-life-examples-of-discriminating-artificial-intelligence-cae395a90070 (accessed 1 November 2021).
42. G.S. Aikenhead and M. Ogawa, Indigenous knowledge and science revisited, *Cultural Studies of Science Education*, 2007, 2, 539–620.
43. BWoY, *The British wheel of yoga*, 2010, http://www.bwy.org.uk/ (accessed 2 March 2022).
44. International Somatic Movement Educators and Therapists Association, http://www.ismeta.org/ (accessed 2 March 2022).
45. B.K. Iyengar, *Light on yoga*, HarperCollins, London, 1966.
46. D. Swenson, *Ashtanga yoga 'The practice manual'*, Ashtanga Yoga Productions, Houaton, TX, 1999.
47. S.K. Pattabhi Jois, *Yoga Mala*, North Point Press, New York, 1999.
48. J. Adler, *Offering from the conscious body: The discipline of authentic movement*, Inner Traditions, Rochester, VT, 2002.
49. L. Hartley, *Wisdom of the body moving*, North Atlantic Books, Berkeley, CA, 1989.
50. L. Hartley, *Somatic psychology: Body, mind and meaning*, Whurr Publishers, London, 2004.
51. J. Leigh, Embodiment and authenticity, in *Lived experiences of ableism in academia: Strategies for inclusion in higher education*, ed N. Brown, Policy Press, Bristol, 2021, pp 53–72.

REFERENCES

52. J. Leigh, Exploring multiple identities: An embodied perspective on academic development and higher education research, *Journal of Dance and Somatic Practices*, 2019, 11(1), 99–114.
53. J. Leigh, Exploring multiple identities: An embodied perspective on academic development and higher education research, *Journal of Dance and Somatic Practices*, 2019, 11(1), 99–114.
54. J. Leigh, An embodied approach in a cognitive discipline, in *Educational futures and fractures*, Palgrave, London, 2019, pp 221–248.
55. J. Leigh, What would a longitudinal rhythmanalysis of a qualitative researcher's life look like? in *Temporality in qualitative inquiry: Theory, methods, and practices*, Routledge, Abingdon, 2020, pp 72–92.
56. H. Lefevbre, *Rhythmanalysis: Space, time and everyday life*, Continuum, London, 2004.
57. D. Lyon, *What is rhythmanalysis?*, Bloomsbury, London, 2019.
58. J. Leigh, Exploring multiple identities: An embodied perspective on academic development and higher education research, *Journal of Dance and Somatic Practices*, 2019, 11(1), 99–114.
59. J. Leigh and N. Brown, Internalised ableism: Of the political and the personal, in *Ableism in academia: Theorising experiences of disabilities and chronic illnesses in higher education*, ed N. Brown and J. Leigh, UCL Press, London, 2020, pp 164–181.
60. Withdrawal: T. Hudlicky, 'Organic synthesis – Where now?' Is thirty years old. A reflection on the current state of affair, https://onlinelibrary.wiley.com/doi/full/10.1002/anie.202006717?af=R (accessed 21 July 2021).
61. Withdrawal: T. Hudlicky, 'Organic synthesis – Where now?' Is thirty years old. A reflection on the current state of affair.
62. N. Compton, T. Kueckmann, F. MaaB, X. Su, S. Tobey and N. Weickgennannt, Angewandte Chemie's redefined international advisory board: Strengthening connections between the journal and its community, *Angewandte Chemie*, 2021, 60(33), 17752–17754.

63. C.A. Urbina-Blanco, S.Z. Jilani, I.R. Speight, M.J. Bojdys, T. Friščić, J.F. Stoddart, T.L. Nelson, J. Mack, R.A.S. Robinson, E.A. Waddell, J.L. Lutkenhaus, M. Godfrey, M.I. Abboud, S.O. Aderinto, D. Aderohunmu, L. Bibič, J. Borges, V.M. Dong, L. Ferrins, F.M. Fung, T. John, F.P.L. Lim, S.L. Masters, D. Mambwe, P. Thordarson, M.-M. Titirici, G.D. Tormet-González, M.M. Unterlass, A. Wadle, V.W.-W. Yam and Y.-W. Yang, A diverse view of science to catalyse change, *Nature Chemistry*, 2020, 12, 773–776.

64. S.C.L. Kamerlin and P. Wittung-Stafshede, Female faculty: Why so few and why care?, *Chemistry: A European Journal*, 2020, 26, 8319–8323.

65. C.J. Burrows, J. Huang, S. Wang, H.J. Kim, G.J. Meyer, K. Schanze, T.R. Lee, J.L. Lutkenhaus, D. Kaplan, C. Jones, C. Bertozzi, L. Kiessling, M.B. Mulcahy, C.W. Lindsley, M.G. Finn, J.D. Blum, P. Kamat, W. Choi, S. Snyder, C.C. Aldrich, S. Rowan, B. Liu, D. Liotta, P.S. Weiss, D. Zhang, K.N. Ganesh, H.A. Atwater, J.J. Gooding, D.T. Allen, C.A. Voigt, J. Sweedler, A. Schepartz, V. Rotello, S. Lecommandoux, S.J. Sturla, S. Hammes-Schiffer, J. Buriak, J.W. Steed, H. Wu, J. Zimmerman, B. Brooks, P. Savage, W. Tolman, T.F. Hofmann, J.F. Brennecke, T.A. Holme, K.M. Merz, G. Scuseria, W. Jorgensen, G.I. Georg, S. Wang, P. Proteau, J.R. Yates, P. Stang, G.C. Walker, M. Hillmyer, L.S. Taylor, T.W. Odom, E. Carreira, K. Rossen, P. Chirik, S.J. Miller, J.-E. Shea, A. McCoy, M. Zanni, G. Hartland, G. Scholes, J.A. Loo, J. Milne, S.B. Tegen, D.T. Kulp and J. Laskin, Confronting racism in chemistry journals, *ACS Applied Materials & Interfaces*, 2020, 12, 28925–28927.

66. S.E. Reisman, R. Sarpong, M.S. Sigman and T.P. Yoon, Organic chemistry: A call to action for diversity and inclusion, *Organic Letters*, 2020, 22, 6223–6228.

67. J. Stockard, J. Greene, G. Richmond and P. Lewis, Is the gender climate in chemistry still chilly? Changes in the last decade and the long-term impact of COACh-sponsored workshops, *Journal of Chemical Education*, 2018, 95, 1492–1499.

68. N. Brown and J. Leigh, *Ableism in academia: Theorising experiences of disabilities and chronic illnesses in higher education*, UCL Press, London, 2020.

69. N. Brown, Ed, *Lived experiences of ableism in academia: Strategies for inclusion in higher education*, Policy Press, Bristol, 2021.

REFERENCES

70. K. Sutherland, Constructions of success in academia: An early career perspective, *Studies in Higher Education*, 2017, 42, 743–759.
71. A. Slater, C. Caltagirone, E. Draper, N. Busschaert, K. Hutchins and L. Leigh, J. Pregnancy in the lab, *Nat Rev Chem*, 2022, 6, 163–164.
72. R. Colwell and S. Bertsch McGrayne, *A lab of one's own: One woman's personal journey through sexism in science*, Simon & Schuster, New York, 2020.
73. S. Rosser, *Breaking into the lab: Engineering progress for women in science*, New York University Press, New York, 2012.
74. E. Daniell, *Every other Thursday: Stories and strategies from successful women scientists*, Yale University Press, New Haven, CT, 2006.
75. A. McConnell, C. Haynes, C. Caltagirone and J. Hiscock, Editorial for the supramolecular chemistry: Young talents and their mentors special collection, *Chempluschem*, 2020, 85, 2544–2545.
76. https://think.taylorandfrancis.com/special_issues/first-gen-supramolecular-chemists/?utm_source=TFO&utm_medium=cms&utm_campaign=JPG15743 (accessed 11 March 2022).
77. A.K. Scaffidi and J.E. Berman, A positive postdoctoral experience is related to quality supervision and career mentoring, collaborations, networking and a nurturing research environment, *Higher Education*, 2011, 62, 685–698.
78. B.M. Cossairt, J.L. Dempsey and E.R. Young, The Chemistry Women Mentorship Network (ChemWMN): A tool for creating critical mass in academic chemistry, *Inorganic Chemistry*, 2019, 58, 12493–12496.
79. H. Laube, Building connections across differences: Faculty mentoring as institutional change, in *Strategies for resisting sexism in the academy: Higher education, gender and intersectionality*, ed G. Crimmins, Palgrave Macmillan, Cham, 2019, pp 95–113.
80. K. Crenshaw, *Demarginalizing the intersection of race and sex: A black feminist critique of antidiscrimination doctrine, feminist theory, and antiracist politics*, University of Chicago Legal Forum, Chicago, 1989.
81. J.S. Leigh, J.R. Hiscock, S. Koops, A.J. McConnell, C.J.E. Haynes, C. Caltagirone, M. Kieffer, A.G. Slater, E.R. Draper, K.M. Hutchins, D. Watkins, N. Busschaert, L.K.S. von Krbek, K.A. Jolliffe and M.J. Hardie, Lived experiences of the supramolecular chemistry community managing their research through COVID-19, *CHEM*, 2022, 8, 299–311.

Chapter Three

1. L. Back, *Academic diary: Or why higher education still matters*, Goldsmiths Press, London, 2016.
2. J. Fanghanel, *Being an academic*, Routledge, Abingdon, 2012.
3. F. Prondzynski, *A university blog: A diary of life and strategy inside and outside the university*, 2014, https://universitydiary.wordpress.com/2014/01/28/do-we-recognise-good-teaching-in-our-universities/#comment-30053 (accessed 2 March 2022).
4. F. Vostal, *Accelerating academia: The changing structure of academic time*, Palgrave Macmillan, Basingstoke, 2016.
5. H. Rosa, Full speed burnout? From the pleasures of the motorcycle to the bleakness of the treadmill: The dual face of social acceleration, *International Journal of Motorcycle Studies*, 2010, 6(1), np.
6. R. Gill, Breaking the silence: The hidden injuries of neo-liberal academia, in *Secrecy and silence in the research process: Feminist reflections*, ed R. Flood and R. Gill, Routledge, London, 2009.
7. S. Acker and C. Armenti, Sleepless in academia, *Gender and Education*, 2004, 16, 3–24.
8. J. Williams, *Consuming higher education: Why education can't be bought*, Bloomsbury, London, 2013.
9. P. Bourdieu, *On television*, The New Press, New York, 1998.
10. R. Murray and D. Mifsud, *The positioning and making of female professors*, Palgrave Macmillan, Cham, 2019.
11. J. Malcolm and M. Zukas, Making a mess of academic work: experience, purpose and identity, *Teaching in Higher Education*, 2009, 14, 495–506.
12. J. Leigh, An embodied approach in a cognitive discipline, in *Educational futures and fractures*, Palgrave, London, 2019.
13. P. Boynton, *Being well in academia*, Routledge, Abingdon, 2020.
14. B. Clark and A. Sousa, *How to be a happy academic*, Sage, London, 2018.
15. K. Kelsky, *The professor is in*, Three Rivers Press, New York, 2015.
16. UK National Account of Wellbeing, *What is wellbeing*, 2012, http://www.nationalaccountsofwellbeing.org/learn/what-is-well-being.html (accessed 22 June 2021).
17. R. Bailey, Wellbeing, happiness and education, *British Journal of Sociology of Education*, 2009, 30, 795–802.
18. J. Griffin, *Well-being*, Clarendon Press, Oxford, 1996.

REFERENCES

19. R. Barrow, *Happiness*, Martin Robinson, Oxford, 1980.
20. D. Parfit, *Reasons and persons*, Clarendon Press, Oxford, 1984.
21. M. Nussbaum, *Women and human development: The capabilities approach*, Cambridge University Press, Cambridge, 2000.
22. A. Sen, *Commodities and capabilities*, Oxford University Press, New Delhi, 1999.
23. R. Dodge, A. Daly, J. Huyton and L. Sanders, The challenge of defining wellbeing, *International Journal of Wellbeing*, 2012, 2, 222–235.
24. K. W. Brown and R. Ryan, The benefits of being present: Mindfulness and its role in psychological well-being, *Journal of Personality and Social Psychology*, 2003, 84, 822–848.
25. J. Adler, *Offering from the conscious body: The discipline of authentic movement*, Inner Traditions, Rochester, VT, 2002.
26. J. Bacon, Authentic movement as well-being practice, in *Oxford handbook for dance and well-being*, ed V. Karkou, S. Lycouris and S. Oliver, Oxford University Press, Oxford, 2015.
27. S. Stiver Lie and V. O'Leary, *Storming the tower: Women in the academic world*, Kogan Page, London, 1990.
28. G. Crimmins, A structural account of inequality in the international academy: Why resistance to sexism remains urgent and necessary, in *Strategies for resisting sexism in the academy: Higher education, gender and intersectionality*, ed G. Crimmins, Palgrave Macmillan, Cham, 2019, pp 3–16.
29. C. Mazak, Babies taught me how to 'do' academia: Crafting a career in an institution that was not built for mothers, in *The positioning and making of female professors: Pushing career advancement open*, ed R. Murray and D. Mifsud, Palgrave Macmillan, Cham, 2019, pp 75–88.
30. E. Yarrow, The power of networks in the gendered academy, in *Women thriving in academia*, ed M. Mahat, Emerald, Bingley, 2021, pp 89–108.
31. S. Rosser, *Academic women in STEM faculty*, Palgrave Macmillan, Basingstoke, 2017.
32. K. Powell, The parenting penalties faced by scientist mothers, *Nature*, 20 July 2021, https://www.nature.com/articles/d41586-021-01993-x?WT.ec_id=NATURE-20210722&utm_source=nature_etoc&utm_medium=email&utm_campaign=20210722&sap-outbound-id=45211DAC4F7B6D10B49010EEB1374001DB14AEE2 (accessed 2 March 2022).

33. E. Evans, Fitting in, in *Mama PhD: Women write about motherhood and academic life*, ed E. Evans and C. Grant, Rutgers University Press, Piscataway, NJ, 2018, pp 49–54.
34. E. Evans and C. Grant, *Mama PhD: Women write about motherhood and academic life*, Rutgers University Press, Piscataway, NJ, 2008.
35. S.A. Shelton, J.E. Flynn and T.J. Grosland, Eds, *Feminism and intersectionality in academia: Women's narratives and experiences in higher education*, Palgrave Macmillan, Basingstoke, 2018.
36. P. Agarwal, *(M)otherhood: On the choices of being a woman*, Canongate Books, Edinburgh, 2021.
37. M.A. Mason and E.M. Ekman, *Mothers on the fast track: How a new generation can balance family and careers*, Oxford University Press, Oxford, 2007.
38. A. Douglass, Three sides of the balance, in *Motherhood, the elephant in the laboratory: Women scientists speak out*, ed E. Monosson, Cornell University Press, Ithaca, NY, 2008.
39. S. Ahmed, *Living a feminist life*, Duke University Press, Durham, NC, 2017.
40. V. Gornick, *Women in science: Then and now*, The Feminist Press, New York, 25th Anniversary, 2009.
41. C. Seroogy, Reflections of a female scientist with outside interests, in *Motherhood, the elephant in the laboratory: Women scientists speak out*, ed E. Monosson, Cornell University Press, Ithaca, NY, 2008.
42. S. Rosser, *Breaking into the lab: Engineering progress for women in science*, New York University Press, New York, 2012.
43. S. Rosser, *The science glass ceiling: Academic women scientists and the struggle to succeed*, Routledge, New York, 2004.
44. R. Colwell and S. Bertsch McGrayne, *A lab of one's own: One woman's personal journey through sexism in science*, Simon & Schuster, New York, 2020.
45. A. Oakley, *From here to maternity: Becoming a mother*, Penguin, Harmondsworth, 1981.
46. A. Brecher, Balancing family and career demands with 20/20 hindsight, in *Motherhood, the elephant in the laboratory: Women scientists speak out*, ed E. Monosson, Cornell University Press, Ithaca, NY, 2008, pp 25–30.

REFERENCES

47. A. P. Abola, Finding my way back to the bench, in *Motherhood, the elephant in the laboratory: Women scientists speak out*, ed E. Monosson, Cornell University Press, Ithaca, NY, 2008, pp 125–129.
48. T. Estermann, E. Bennetot Pruvot, V. Kupriyanova and H. Stoyanova, *The impact of the Covid-19 crisis on university funding in Europe: Lessons learnt from the 2008 global financial crisis*, European University Association, Geneva, 2020.
49. Anon, 'I have been cowed into silence': An academic speaks out against redundancy, 2020, https://universitybusiness.co.uk/comment/covid-19-redundancies-in-higher-education-an-academic-speaks-out/ (accessed 2 March 2022).
50. J. Butler and G. Yancy, *Mourning is a political act amid the pandemic and its disparities*, 2020, https://discoversociety.org/2020/04/10/covid-19-what-it-means-to-think-violently/ (accessed 2 March 2022).
51. I. Dougall, M. Weick and M. Vasiljevic, *Inside UK universities: Staff mental health and wellbeing during the coronavirus pandemic*, Durham University, Durham, 2021.
52. C. Flaherty, Inside higher education, www.insidehighered.com/news/2020/04/21/early-journal-submission-data-suggest-covid-19-tanking-womens-research-productivity (accessed 10 November 2020).
53. J.S. Leigh, J.R. Hiscock, S. Koops, A.J. McConnell, C.J.E. Haynes, C. Caltagirone, M. Kieffer, A.G. Slater, E.R. Draper, K.M. Hutchins, D. Watkins, N. Busschaert, L.K.S. von Krbek, K.A. Jolliffe and M.J. Hardie, Lived experiences of the supramolecular chemistry community managing their research through COVID-19, *CHEM*, 2022, 8, 299–311.
54. J. Gans, *Joshua Gans*, 2020, https://www.joshuagans.com/economics-in-the-age-of-covid19 (accessed 2 March 2022).
55. J. Blanden, C. Crawford, L. Fumagalli and B. Rabe, *School closures and parents' mental health*, Institute for Social and Economic Research, University of Essex, 2021.
56. B.P. Gabster, K. van Daalen, R. Dhatt and M. Barry, Challenges for the female academic during the COVID-19 pandemic, *Lancet (London, England)*, 2020, 395, 1968–1970.
57. A. O'Reilly and F.J. Green, Eds, *Mothers, mothering, and COVID-19: Dispatches from a pandemic*, Demeter, Bradford, Ontario, 2021.

58. K. 'Fire' Kovarovic, M. Dixon, K. Hall and N. Westmarland, *The impact of Covid-19 on mothers working in UK higher education institutions*, Durham University, Durham, 2021.
59. K. Kramer, *Chemistry World*, 2020, https://www.chemistryworld.com/news/uk-chemistry-pipeline-loses-almost-all-of-its-black-asian-and-other-ethnic-minority-chemists-after-undergraduate-studies/4012258.article#/ (accessed 2 March 2022).
60. L. Svinth, Leaky pipeline – to be or not to be a useful metaphor in understanding why women to a disproportional degree exit from scientific careers, *The Danish University of Education*, https://www.semanticscholar.org/paper/Leaky-pipeline-"---to-be-or-not-to-be-a-useful-in-a-Svinth/866433f8414924ac699733dcf1047700d3818c39 (accessed 2 March 2022).
61. B. Cornell, *PhD students and their careers*, HEPI, London, 2021.
62. RSC, *Breaking the barriers: Women's retention and progression in the chemical sciences*, Royal Society of Chemistry, London, 2019.
63. E. Daniell, *Every other Thursday: Stories and strategies from successful women scientists*, Yale University Press, New Haven, CT, 2006.
64. D. Kalekin-Fishman, Mis-making an academic career: Power, discipline, structures, and practices, in *The positioning and making of female professors: Pushing career advancement open*, ed R. Murray and D. Mifsud, Palgrave Macmillan, Cham, 2019, pp 177–200.

Chapter Four

1. I. Oluo, *So you want to talk about race*, Hatchett Press Group, New York, 2020.
2. R. Eddo-Lodge, *Why I'm no longer talking to white people about race*, Bloomsbury, London, 2017.
3. MaNishtana, *Thoughts from a unicorn: 100% Black. 100% Jewish. 0% safe.*, Multikosheral Press, New York, 2012.
4. L. Saad, *Me and white supremacy: How to recognise your privilege, combat racism and change the world*, Quercus, London, 2020.
5. A. Lorde, *The master's tools will never dismantle the master's house*, Penguin Random House, Milton Keynes, 2017.

REFERENCES

6. S. Ahmed, *On being included: Racism and diversity in institutional life*, Duke University Press, Durham, NC, 2012.
7. S. Fryberg and E. Martinez, Eds, *The truly diverse faculty: New dialogues in American higher education*, Palgrave Macmillan, Basingstoke, 2014.
8. W.H. McGee and E.O. Robinson, Eds, *Diversifying STEM: Multidisciplinary perspectives on race and gender*, Rutgers University Press, New Brunswick, NJ, 2020.
9. W.H. Robinson, E.O. McGee, L.C. Bentley, S.L. Houston and P.K. Botchway, Addressing negative racial and gendered experiences that discourage academic careers in engineering, *Computing in Science & Engineering*, 2016, 18(2), 29–39.
10. B. Yoder, *Engineering by the numbers*, American Society for Engineering Education, Washington, DC, 2015.
11. National Science Foundation, https://nsf.gov/statistics/reports.cfm (accessed 1 November 2021).
12. R. Merton, *On social structure and science*, University of Chicago Press, Chicago, IL, 1996.
13. L. Dubois Baber, Color-blind liberalism in postsecondary STEM education, in *Diversifying STEM: Multidisciplinary perspectives on race and gender*, ed E. McGee and W. Robinson, Rutgers University Press, New Brunswick, NJ, 2020, pp 19–35.
14. V. Gornick, *Women in science: Then and now*, The Feminist Press, New York, 25th Anniversary, 2009.
15. HESA, Academic staff by ethnicity and occupation 2019/2020, www.hesa.ac.uk/data-and-analysis/staff/working-in-he/characteristics#acempfunchar (accessed 2 July 2021).
16. Gov, Population of the UK, www.ethnicity-facts-figures.service.gov.uk/uk-population-by-ethnicity/national-and-regional-populations/population-of-england-and-wales/latest (accessed 2 July 2021).
17. S. Coughlan, Only 1% of UK university professors are black, *BBC*, 19 January 2021, https://www.bbc.co.uk/news/education-55723120 (accessed 2 March 2022).
18. A. Vaughan, Only 10 senior Black researchers awarded UK science funding last year, www.newscientist.com/article/2262849-only-10-senior-black-researchers-awarded-uk-science-funding-last-year/ (accessed 2 July 2021).

19. A. Prasad, Why are there still so few Black scientists in the UK, *The Observer*, 10 April 2021, https://www.theguardian.com/science/2021/apr/10/why-are-there-still-so-few-black-scientists-in-the-uk (accessed 2 March 2022).
20. M. Makgoba, Black scientists matter, *Science (80-)*, 2020, 369, 884.
21. K. Laland, Racism in academia, and why the 'little things' matter, *Nature*, 2020, 584, 653–654.
22. T. Heffernan, Sexism, racism, prejudice, and bias: A literature review and synthesis of research surrounding student evaluations of courses and teaching, *Assessment & Evaluation in Higher Education*, 2021, 47(1), 144–154.
23. L. Ackerman-Biegasiewicz, D. Rias-Rotondo and D. Biegasiewicz, Organic chemistry: A retrosynthetic approach to a diverse field, *ACS Central Science*, 2020, 6, 1845–1850.
24. B. Menon, The missing colours of chemistry, *Nature Chemistry*, 2021, 13, 101–106.
25. S.E. Reisman, R. Sarpong, M.S. Sigman and T.P. Yoon, Organic chemistry: A call to action for diversity and inclusion, *Organic Letters*, 2020, 22, 6223–6228.
26. C.A. Urbina-Blanco, S.Z. Jilani, I.R. Speight, M.J. Bojdys, T. Friščić, J.F. Stoddart, T.L. Nelson, J. Mack, R.A.S. Robinson, E.A. Waddell, J.L. Lutkenhaus, M. Godfrey, M.I. Abboud, S.O. Aderinto, D. Aderohunmu, L. Bibič, J. Borges, V.M. Dong, L. Ferrins, F.M. Fung, T. John, F.P.L. Lim, S. L. Masters, D. Mambwe, P. Thordarson, M.-M. Titirici, G.D. Tormet-González, M.M. Unterlass, A. Wadle, V.W.-W. Yam and Y.-W. Yang, A diverse view of science to catalyse change, *Nature Chemistry*, 2020, 12, 773–776.
27. RSC, *Diversity landscape of the chemical sciences*, Royal Society of Chemistry, London, 2018.
28. C.J. Burrows, J. Huang, S. Wang, H.J. Kim, G.J. Meyer, K. Schanze, T.R. Lee, J.L. Lutkenhaus, D. Kaplan, C. Jones, C. Bertozzi, L. Kiessling, M.B. Mulcahy, C.W. Lindsley, M.G. Finn, J.D. Blum, P. Kamat, W. Choi, S. Snyder, C.C. Aldrich, S. Rowan, B. Liu, D. Liotta, P.S. Weiss, D. Zhang, K. N. Ganesh, H.A. Atwater, J.J. Gooding, D.T. Allen, C.A. Voigt, J. Sweedler, A. Schepartz, V. Rotello, S. Lecommandoux, S.J. Sturla, S. Hammes-Schiffer, J. Buriak, J.W. Steed, H. Wu, J. Zimmerman, B.

REFERENCES

Brooks, P. Savage, W. Tolman, T.F. Hofmann, J.F. Brennecke, T.A. Holme, K.M. Merz, G. Scuseria, W. Jorgensen, G.I. Georg, S. Wang, P. Proteau, J.R. Yates, P. Stang, G.C. Walker, M. Hillmyer, L.S. Taylor, T.W. Odom, E. Carreira, K. Rossen, P. Chirik, S.J. Miller, J.-E. Shea, A. McCoy, M. Zanni, G. Hartland, G. Scholes, J.A. Loo, J. Milne, S.B. Tegen, D.T. Kulp and J. Laskin, Confronting racism in chemistry journals, *ACS Applied Materials & Interfaces*, 2020, 12, 28925–28927.

29. K. Kramer, UK chemistry pipeline loses almost all of its BAME students after undergraduate studies, *Chemistry World*, 10 August 2020, https://www.chemistryworld.com/news/uk-chemistry-pipeline-loses-almost-all-of-its-black-asian-and-other-ethnic-minority-chemists-after-undergraduate-studies/4012258.article#/ (accessed 2 March 2022).

30. S. Iyer, D. Stallings and R. Hernandez, *National Diversity Equity Workshop 2011: Lowering barriers for all underrepresented chemistry professors*, American Chemical Society, Washington, DC, 2018.

31. K. Crenshaw, *Demarginalizing the intersection of race and sex: A black feminist critique of antidiscrimination doctrine, feminist theory, and antiracist politics*, University of Chicago Legal Forum, Chicago, IL, 139–167, 1989.

32. M. Cox, in *Diversifying STEM: Multidisciplinary perspectives on race and gender2*, Rutgers University Press, New Brunswick, NJ, 2020, pp 53–66.

33. W. Joice and A. Tetlow, *Disability STEM data for students and academic staff in higher education 2007/08 to 2018/19*, Royal Society, London, 2021.

34. GOV UK, *UK Government Legislation*, 2010, http://www.legislation.gov.uk/ukpga/2010/15/contents (accessed 2 March 2022).

35. N. Brown and J. Leigh, Ableism in academia: Where are the disabled and ill academics?, *Disability & Society*, 2018, 33(6), 985–989.

36. HESA, *Who's working in HE?*, 2020, https://www.hesa.ac.uk/data-and-analysis/staff/working-in-he (accessed 2 March 2022).

37. CRAC, *Qualitative research on barriers to progression of disabled scientists*, Royal Society, London, 2020.

38. N. Brown and J. Leigh, *Ableism in academia: Theorising experiences of disabilities and chronic illnesses in higher education*, UCL Press, London, 2020.

39. N. Brown, Ed, *Lived experiences of ableism in academia: Strategies for inclusion in higher education*, Policy Press, Bristol, 2021.
40. J. Leigh and N. Brown, Internalised ableism: Of the political and the personal, in *Ableism in academia: Theorising experiences of disabilities and chronic illnesses in higher education*, ed N. Brown and J. Leigh, UCL Press, London, 2020, pp 164–181.
41. C. Finesilver, J. Leigh and N. Brown, Invisible disability, unacknowledged diversity, in *Ableism in academia: Theorising experiences of disabilities and chronic illnesses in higher education*, ed N. Brown and J. Leigh, UCL Press, London, 2020, pp 143–160.
42. D. Turner, *Super humans or scroungers*, 2012, http://www.historyandpolicy.org/opinion-articles/articles/superhumans-or-scroungers-disability-past-and-present (accessed 2 March 2022).
43. K. Rummery, From the personal to the political: Ableism, activism and academia, in *Ableism in academia: Theorising experiences of disabilities and chronic illnesses in higher education*, ed N. Brown and J. Leigh, UCL Press, London, 2020, pp 182–201.
44. L. Harris, *Exploring the effect of disability microaggressions on sense of belonging and participation in college classrooms*, School of Psychology, Utah State University, Logan, Utah, 2017.
45. M. Leicester and T. Lovell, Disability voice: Educational experience and disability, *Disability & Society*, 1997, 12, 111–118.
46. A. Mackleden, *Being disabled is exhausting*, 2019, https://clarissaexplainsfa.com/blog/2019/2/1/being-disabled-is-exhausting (accessed 2 March 2022).
47. J. Brock, 'Textbook case' of disability discrimination in grant applications, *Nature Index*, January 2021, https://www.natureindex.com/news-blog/textbook-case-of-disability-discrimination-in-research-grant-applications (accessed 2 March 2022).
48. M. Leonardi, J. Bickenbach, T. Ustun, N. Kostanjsek and S. Chatterji, The definition of disability: what is in a name?, *Lancet*, 2006, 368, 1219–1221.
49. M. Oliver, The social model of disability: Thirty years on, *Disability & Society*, 2013, 28, 1024–1026.

REFERENCES

50. R. Colwell and S. Bertsch McGrayne, *A lab of one's own: One woman's personal journey through sexism in science*, Simon & Schuster, New York, 2020.
51. M. Shildrick, *Leaky bodies and boundaries: Feminism, postmodernism and (bio)ethics*, Routledge, Abingdon, 1997.
52. Aristotle, *De natura animalium, De partibus animalium, De generatione animalium*, Venetiis: Impressum mandato & expensis nobilis uiri Domini Octauiani Scoti ciuis Modoeti esis per Bartholameum de Zanis de Portesio, Venice.
53. Galen, *de usu partium*, Cornell University Press, Ithaca, NY, 1968.
54. P. Fara, *A lab of one's own: Science and suffrage in the first world war*, Oxford University Press, Oxford, 2018.
55. E. Daniell, *Every other Thursday: Stories and strategies from successful women scientists*, Yale University Press, New Haven, CT, 2006.
56. S. Rosser, *Breaking into the lab: Engineering progress for women in science*, New York University Press, New York, 2012.
57. Y. Xie and K. Shauman, *Women in science*, Harvard University Press, Boston, MA, 2003.
58. HESA, HE student enrolments by subject area and sex 2014/15 to 2018/19, 2019, https://www.hesa.ac.uk/data-and-analysis/sb255/figure-13 (accessed 1 March 2022).
59. S. Rosser, *Academic women in STEM faculty*, Palgrave Macmillan, Basingstoke, 2017.
60. Women in STEM, Women in STEM statistics – General outlook for female students, https://www.stemwomen.com/blog/2021/01/women-in-stem-percentages-of-women-in-stem-statistics#:~:text=On%20 the%20surface%20this%20would,up%20to%2026%25%20in%202 018 (accessed 2 March 2022).
61. HESA, HE student enrolments, www.hesa.ac.uk/data-and-analysis/sb255/figure-13 (accessed 11 November 2020).
62. B. Thege, S. Popescue-Willigmann, R. Pioch and S. Badri-Hoher, Eds, *Paths to career and success for women in science: Findings from international research*, Springer VS, Wiesbaden, 2014.
63. M.L. Dion, J.L. Sumner and S.M. Mitchell, Gendered citation patterns across political science and social science methodology fields, *Political Analysis*, 2018, 26, 312–327.

64. F. Squazzoni, G. Bravo, P. Dondio, M. Farjam, A. Marusic, B. Mehmani, M. Willis, A. Birukou and F. Grimaldo, No evidence of any systemic bias against manuscripts by women in the peer review process of 145 scholarly journals, *SocArXiv*, 2020, https://osf.io/preprints/socarxiv/gh4rv/ (accessed 2 March 2022).
65. K. Hyland and F. Jiang, 'This work is antithetical to the spirit of research': An anatomy of harsh peer reviews, *Journal for English for Academic Purposes*, 2020, 46, https://doi.org/10.1016/j.jeap.2020.100867.
66. G. Bravo, F. Grimaldo, E. Lopez-Inesta, B. Mehmani and F. Squazzoni, The effect of publishing peer review reports on referee behavior in five scholarly journals, *Nature Communications*, 2019, 10, 322.
67. RSC, *Breaking the barriers: Women's retention and progression in the chemical sciences*, Royal Society of Chemistry, London, 2019.
68. RSC, *Is publishing in the chemical sciences gender biased? Driving change in research culture*, Royal Society of Chemistry, London, 2019.
69. NSF, ADVANCE: Organizational Change for Gender Equity in STEM Academic Professions (ADVANCE), www.nsf.gov/funding/pgm_summ.jsp?pims_id=5383 (accessed 24 July 2021).
70. M. Tsouroufli, An examination of the Athena SWAN initiatives in the UK: A critical reflection, in *Strategies for resisting sexism in the academy: Higher education, gender and intersectionality*, ed G. Crimmins, Palgrave Macmillan, Chaim, 2019, pp 35–54.
71. K. Campion and K. Clark, Revitalising race equality policy? Assessing the impact of the Race Equality Charter mark for British universities, *Race, Ethnicity and Education*, 2021, 25(1), 18–37.
72. M.A. Mason and E.M. Ekman, *Mothers on the fast track: How a new generation can balance family and careers*, Oxford University Press, Oxford, 2007.
73. S. Rosser, *Feminism within the science and health care professions: Overcoming resistance*, Pergamon Press, Oxford, 1988.
74. S. Rosser, *The science glass ceiling: Academic women scientists and the struggle to succeed*, Routledge, New York, 2004.
75. E. Yarrow, The power of networks in the gendered academy, in *Women thriving in academia*, ed M. Mahat, Emerald, Bingley, 2021, pp 89–108.

REFERENCES

76. M. Mahat, R. Hardiman, K. Howell and I. Mateo-Babiano, Women leading in academia, in *Women thriving in academia*, ed M. Mahat, Emerald, Bingley, 2021, pp 51–69.
77. R. Murray and D. Mifsud, *The positioning and making of female professors*, Palgrave Macmillan, Cham, 2019.
78. S.A. Shelton, J.E. Flynn and T.J. Grosland, Eds, *Feminism and intersectionality in academia: Women's narratives and experiences in higher education*, Palgrave Macmillan, Basingstoke, 2018.
79. Intrac, Action learning sets: A guide for small and diaspora NGOs, www.intrac.org/wpcms/wp-content/uploads/2016/09/Action-Learning-Sets-An-INTRAC-guide-1.pdf (accessed 30 June 2021).

Chapter Five

1. J.M. Lehn, Toward complex matter: Supramolecular chemistry and self-organization, *Proceedings of the National Academy of Sciences*, 2002, 99, 4763–4768.
2. Special issue honouring Professor Rocco Ungaro, *Supramolecular Chemistry*, www.tandfonline.com/toc/gsch20/25/9-11?nav=tocList (accessed 21 July 2021).
3. ISMSC, ISMSC prizes, www.chem.byu.edu/alumni/ismsc/awards/ (accessed 21 July 2021).
4. J. Leigh, What would a longitudinal rhythmanalysis of a qualitative researcher's life look like? in *Temporality in qualitative inquiry: Theory, methods, and practices*, Routledge, Abingdon, 2020.
5. J. Leigh and N. Brown, *Embodied inquiry: Research methods*, Bloomsbury, London, 2021.
6. V. Braun and V. Clarke, To saturate or not to saturate? Questioning data saturation as a useful concept for thematic analysis and sample-size rationales, *Qualitative Research in Sport, Exercise and Health*, 2021, 13, 201–216.
7. N. Brown and J. Leigh, Ableism in academia: where are the disabled and ill academics?, *Disability & Society*, 2018, 33(6), 985–989.
8. N. Brown and J. Leigh, *Ableism in academia: Theorising experiences of disabilities and chronic illnesses in higher education*, UCL Press, London, 2020.

9. J. Hiscock and J. Leigh, Exploring perceptions of and supporting dyslexia in teachers in higher education in STEM, *Innovations in Education and Teaching International*, 2020, 57, 714–723.
10. J. Hiscock and J. Leigh, Teaching with and supporting teachers with dyslexia in higher education, in *Lived experiences of ableism in academia: Strategies for inclusion in higher education*, ed N. Brown, Policy Press, Bristol, 2021.
11. NADSN, NADSN, www.nadsn-uk.org (accessed 7 June 2021).
12. J. Sarju, Nothing about us without us – Towards genuine inclusion of disabled scientists and science students post pandemic, *Chemistry: A European Journal*, 2021, 27(41), 10489–10494.
13. A. Standlee, *On the borders of the academy: Challenges and strategies for first-generation graduate students and faculty*, Syracuse University Press, Syracuse, NY, 2018.
14. P. Bourdieu, *Outline of a theory of practice*, Cambridge University Press, Cambridge, 1977.
15. P. Bourdieu R., *The logic of practice*, Stanford University Press, Stanford, CA, 1990.
16. C. Caltagirone, E. Draper, C. Haynes, M. Hardie, J. Hiscock, K. Jolliffe, M. Kieffer, A. McConnell and J. Leigh, An area specific, international community-led approach to understanding and addressing EDI issues within supramolecular chemistry, *Angewandte Chemie International Edition*, 2021, **60**, 11572–11579.
17. S. Ahmed, *On being included: Racism and diversity in institutional life*, Duke University Press, Durham, NC, 2012.
18. A. Prasad, Why are there still so few Black scientists in the UK, *The Observer*, 10 April 2021, https://www.theguardian.com/science/2021/apr/10/why-are-there-still-so-few-black-scientists-in-the-uk (accessed 2 March 2022).
19. P. Chakravartty, R. Kuo, V. Grubbs and C. McIlwain, #CommunicationSoWhite, *Journal of Communication*, 2018, 68, 254–266.
20. S. Ahmed, *Living a feminist life*, Duke University Press, Durham, NC, 2017.
21. S. Ahmed, *Complaint!*, Duke University Press, Durham, NC, 2021.

REFERENCES

22. B. Bjørkelo, Workplace bullying after whistleblowing: future research and implications, *Journal of Managerial Psychology*, 2013, 28, 306–323.
23. A. Witze, Sexual harassment is rife in the sciences, finds landmark US study, *Nature*, 2018, https://www.nature.com/articles/d41586-018-05404-6 (accessed 2 March 2022).
24. K.B.H. Clancy, L.M. Cortina and A.R. Kirkland, Opinion: Use science to stop sexual harassment in higher education, *Proceedings of the National Academy of Sciences*, 2020, 202016164.
25. N.C. Cantalupo, And even more of us are brave: intersectionality & sexual harassment of women students of color, *Harvard Womens' Law Journal*, 2019, 42, 1.
26. C. Caltagirone, E. Draper, M. Hardie, C. Haynes, J. Hiscock, K. Jolliffe, M. Kieffer, J. Leigh and A. McConnell, Calling in support, *Chemistry World*, March 2021, https://www.chemistryworld.com/opinion/supramolecular-community-calls-in-support-for-gender-equity/4013248.article (accessed 2 March 2022).
27. @jonathan_steed, WISC are putting my generation to shame, Twitter, https://twitter.com/jonathan_steed/status/1371931380242862087 (accessed 13 July 2021).
28. RSC, Macrocycllic and Supramolecular Chemistry Special Interest Group, www.rsc.org/membership-and-community/connect-with-others/through-interests/interest-groups/masc/ (accessed 16 July 2021).
29. A. Lorde, *The master's tools will never dismantle the master's house*, Penguin Random House, Milton Keynes, 2017.
30. S. Franklin, sexism as a means of reproduction, *New Form.*, 2015, 86, 14–33.
31. WISC, *WISC 1st Skills Workshop*, 2021, https://d017147e-00c2-4e9e-88fd-affbf75f13ce.filesusr.com/ugd/e3c05f_ef7a72f4eaeb4dc58f83e3c84dee4f36.pdf (accessed 2 March 2022).
32. R.A. Dawood and A.-J. Avestro, A new equilibrium for supramolecular chemists, *Nature Chemistry*, 2021, 13, 1164–1165.
33. A. McConnell, C. Haynes, C. Caltagirone and J. Hiscock, Editorial for the Supramolecular Chemistry: Young talents and their mentors special collection, *Chempluschem*, 2020, 85, 2544–2545.

34. C. Caltagirone, E. Draper, M. Hardie, C. Haynes, J. Hiscock, K. Jolliffe, M. Kieffer, J. Leigh and A. McConnell, Calling in support, *Chemistry World*, March 2021, https://www.chemistryworld.com/opinion/supramolecular-community-calls-in-support-for-gender-equity/4013248.article (accessed 2 March 2022).
35. J.S. Leigh, J.R. Hiscock, S. Koops, A.J. McConnell, C.J.E. Haynes, C. Caltagirone, M. Kieffer, A.G. Slater, E.R. Draper, K.M. Hutchins, D. Watkins, N. Busschaert, L.K.S. von Krbek, K.A. Jolliffe and M.J. Hardie, Lived experiences of the supramolecular chemistry community managing their research through COVID-19, *CHEM*, 2022, 8, 299–311.
36. J. Leigh, J. Hiscock, A. McConnell, C. Haynes, M. Kieffer, C. Caltagirone, K. Hutchins, D. Watkins, A. Slater, L. von Krbek, E. Draper and N. Busschaert. WISC – narratives of resilience and community building in a gender constrained field, in *Women in academia*, ed M. Ronksley-Pavia, M. Neumann, J. Manakil and K. Pickard-Smith, Bloomsbury, London.
37. EFeMS, Empowering female minds in STEM: Showing African women that their possibilities are endless, www.empoweringfems.com (accessed 16 July 2021).
38. BA, British Academy APEX Awards, www.thebritishacademy.ac.uk/funding/apex-awards/ (accessed 24 July 2021).

Chapter Six

1. E. Daniell, *Every other Thursday: Stories and strategies from successful women scientists*, Yale University Press, New Haven, CT, 2006.
2. E. Yarrow, The power of networks in the gendered academy, in *Women thriving in academia*, ed M. Mahat, Emerald, Bingley, 2021, pp 89–108.
3. RSC, *Diversity landscape of the chemical sciences*, Royal Society of Chemistry, London, 2018.
4. CRAC, *Qualitative research on barriers to progression of disabled scientists*, Royal Society, London, 2020.
5. UKRI, *Diversity results for UKRI funds*, UKRI, London, 2020.
6. C. Flaherty, Inside higher education, www.insidehighered.com/news/2020/04/21/early-journal-submission-data-suggest-covid-19-tanking-womens-research-productivity (accessed 10 November 2020).

REFERENCES

7. B.P. Gabster, K. van Daalen, R. Dhatt and M. Barry, Challenges for the female academic during the COVID-19 pandemic, *Lancet (London, England)*, 2020, 395, 1968–1970.
8. V. O'Leary and J. Mitchell, Women connecting with women: Networks and mentors, in *Storming the tower: Women in the academic world*, ed S. Stiver Lie and V. O'Leary, Kogan Page, London, 1990, pp 58–73.
9. C. Caltagirone, E. Draper, C. Haynes, M. Hardie, J. Hiscock, K. Jolliffe, M. Kieffer, A. McConnell and J. Leigh, An area specific, international community-led approach to understanding and addressing EDI issues within supramolecular chemistry, *Angewandte Chemie International Edition*, 2021, 60, 11572–11579.
10. A. Bleakley, From reflective practice to holisitic reflexivity, *Studies in Higher Education*, 1999, 24, 315–330.
11. J.S. Leigh, J.R. Hiscock, S. Koops, A.J. McConnell, C.J.E. Haynes, C. Caltagirone, M. Kieffer, A.G. Slater, E.R. Draper, K.M. Hutchins, D. Watkins, N. Busschaert, L.K.S. von Krbek, K.A. Jolliffe and M.J. Hardie, Managing research throughout COVID-19: Lived experiences of supramolecular chemists, *CHEM*, 2022, 8, 299–311.
12. Gov.UK, Coronavirus (COVID-19) in the UK, https://coronavirus.data.gov.uk (accessed 20 July 2021).
13. H. Gye and J. Merrick, Creator of Oxford vaccine Sarah Gilbert warns vulnerable will have to shield as Covid-19 cases rise, *iNews*, 9 July 2021, https://inews.co.uk/news/politics/creator-oxford-vaccine-sarah-gilbert-warns-vulnerable-will-have-shield-covid-19-cases-rise-1096315 (accessed 2 March 2022).
14. Welsh Government, *Locked out: Liberating disabled people's lives and rights in Wales beyond COVID-19*, Welsh Government, Cardiff, 2021.
15. @tnewtondunn, New guidance is coming from Dept of Health in a few days for the immuno-suppressed and clinically very vulnerable, https://twitter.com/tnewtondunn/status/1412904748362182661 (accessed 12 July 2021).
16. M. Ethington, Women tending to their basic needs is not self-care, *Forge*, January 2021, https://forge.medium.com/im-tired-of-basic-human-needs-being-seen-as-self-care-for-women-54ef206e918c (accessed 2 March 2022).

17. A. O'Reilly and F. J. Green, Eds, *Mothers, mothering, and COVID-19: Dispatches from a pandemic*, Demeter, Bradford, Ontario, 2021.
18. J. Blanden, C. Crawford, L. Fumagalli and B. Rabe, *School closures and parents' mental health*, Institute for Social and Economic Research, University of Essex, 2021.
19. C. Prince, Women's witness, in *Mothers, mothering, and COVID-19: Dispatches from the pandemic*, ed A. O'Reilly and F. J. Green, Demeter, Bradford, Ontario, 2021, pp 37–39.
20. A. Abbott, COVID's mental-health toll: How scientists are tracking a surge in depression, *Nature*, 3 February 2021, https://www.nature.com/articles/d41586-021-00175-z (accessed 2 March 2022).
21. I. Dougall, M. Weick and M. Vasiljevic, *Inside UK universities: Staff mental health and wellbeing during the coronavirus pandemic*, Durham University, Durham, 2021.

Chapter Seven

1. J.S. Leigh, J.R. Hiscock, S. Koops, A.J. McConnell, C.J.E. Haynes, C. Caltagirone, M. Kieffer, A.G. Slater, E.R. Draper, K.M. Hutchins, D. Watkins, N. Busschaert, L.K.S. von Krbek, K.A. Jolliffe and M.J. Hardie, Managing research throughout COVID-19: Lived experiences of supramolecular chemists, *CHEM*, 2022, 8, 299–311.
2. RSC, *Breaking the barriers: Women's retention and progression in the chemical sciences*, Royal Society of Chemistry, London, 2019.
3. CRAC, *Qualitative research on barriers to progression of disabled scientists*, Royal Society of Chemistry, London, 2020.
4. RSC, *Diversity landscape of the chemical sciences*, Royal Society of Chemistry, London, 2018.
5. E. Yarrow, The power of networks in the gendered academy, in *Women thriving in academia*, ed M. Mahat, Emerald, Bingley, 2021, pp 89–108.
6. E. Daniell, *Every other Thursday: Stories and strategies from successful women scientists*, Yale University Press, New Haven, CT, 2006.
7. M. Mahat, R. Hardiman, K. Howell and I. Mateo-Babiano, Women leading in academia, in *Women thriving in academia*, ed M. Mahat, Emerald, Bingley, 2021, pp 51–69.

REFERENCES

8. P. Agarwal, *(M)otherhood: On the choices of being a woman*, Canongate Books, Edinburgh, 2021.
9. D. Macht and D. Lubin, A phyto-pharmological study of menstrual toxin, *Journal of Pharmacollogy and Experimental Therapeutics*, 1923, 22, 413–466.
10. N. Brown and J. Leigh, *Ableism in academia: Theorising experiences of disabilities and chronic illnesses in higher education*, UCL Press, London, 2020.
11. N. Brown, Ed, *Lived experiences of ableism in academia: Strategies for inclusion in higher education*, Policy Press, Bristol, 2021.
12. S. Ahmed, *On being included: Racism and diversity in institutional life*, Duke University Press, Durham, NC, 2012.
13. N. Brown and J. Leigh, Ableism in academia: where are the disabled and ill academics?, *Disability & Society*, 2018, 33(6), 985–989.
14. C. Criado Perez, *Invisible women: Exposing data bias in a world designed for men*, Chatto & Windus, London, 2019.
15. A. Grinstein and R. Treister. The unhappy postdoc: a survey based study, *F1000Research*, 2018, 6(1), 1642.
16. N. Roll, Calling attention to a postdoc's struggles and suicide, www.insidehighered.com/news/2017/08/08/scientific-papers-acknowledgments-section-calls-reform-postdocs-treatment (accessed 13 October 2020).
17. A.K. Scaffidi and J.E. Berman, A positive postdoctoral experience is related to quality supervision and career mentoring, collaborations, networking and a nurturing research environment, *Higher Education*, 2011, 62, 685–698.
18. S. Xuhong, The impacts of postdoctoral training on scientists' academic employment, *Journal of Higher Education*, 2013, 84, 239–265.
19. H. Sauermann and M. Roach, Why pursue the postdoc path? Scientific workforce, *Science (80-)*, 2016, 352, 663–664.
20. K. Powell, The future of the postdoc, *Nature*, 2015, 520, 144–147.
21. Vitae, *5 steps forward report*, 2017, https://www.vitae.ac.uk/vitae-publications/reports/vitae-5-steps-forward-web.pdf (accessed 2 March 2022).
22. A.L. Goodwin, Learning to lead by saying yes, in *Women thriving in academia*, ed M. Mahat, Emerald, Bingley, 2021, pp 71–85.

23. V. Evans-Winters, *Black feminism in qualitative inquiry*, Routledge, Abingdon, 2019.
24. J. Leigh and N. Brown, *Embodied inquiry: Research methods*, Bloomsbury, London, 2021.
25. K. Crenshaw, *Demarginalizing the intersection of race and sex: A black feminist critique of antidiscrimination doctrine, feminist theory, and antiracist politics*, University of Chicago Legal Forum, Chicago, IL, 1989.
26. Cambridge Dictionary, Crafting, https://dictionary.cambridge.org/dictionary/english/crafting (accessed 14 July 2021).
27. E. Monosson, *Motherhood, the elephant in the laboratory: Women scientists speak out*, Cornell University Press, Ithaca, NY, 2008.
28. J. Leigh, What would a longitudinal rhythmanalysis of a qualitative researcher's life look like? in *Temporality in qualitative inquiry: Theory, methods, and practices*, Routledge, Abingdon, 2020.
29. C. Caltagirone, E. Draper, C. Haynes, M. Hardie, J. Hiscock, K. Jolliffe, M. Kieffer, A. McConnell and J. Leigh, An area specific, international community-led approach to understanding and addressing EDI issues within supramolecular chemistry, *Angewandte Chemie International Edition*, 2021, 60, 11572–11579.
30. C. Caltagirone, E. Draper, M. Hardie, C. Haynes, J. Hiscock, K. Jolliffe, M. Kieffer, J. Leigh and A. McConnell, Calling in support, *Chemistry World*, March 2021, https://www.chemistryworld.com/opinion/supramolecular-community-calls-in-support-for-gender-equity/4013248.article (accessed 2 March 2022).

Index

#BlackLivesMatter *see* racism
#EveryonesInvited 4
#MeToo 4

A

ableism 5, 27, 56, 73, 75
  internalised 5
Ahmed, Sara 3
Athena Swan 62
autoethnography
  collaborative
    autoethnography 20, 38–39,
      43, 103, 106, 110–113
  as method 16–18, 21–22, 85,
    90–91, 109–110
  see also Embodied Inquiry

B

bias, in academia 13, 52, 120
  in publishing 52
  in recruitment 31
  research on 53, 60, 63
  in science 6, 23, 47, 58–60
Black academics 50, 53–55,
  86–87, 110
  see also racism
Black feminism 2, 18, 55, 77
Bourdieu, Pierre 35
  see also capital
Brecher, Aviva 3, 43
Burke, Tarana *see* #MeToo
Busschaert, Nathalie 73, 87

C

called out 14, 75–76
calling in 32, 14, 17, 75
Caltagirone, Claudia 71, 81
capital 54, 73–74
career 42, 48, 121
challenge, challenging cultural
  norms 26, 52, 62, 75–77,
    119, 123
  in science 20, 27, 33, 47, 53–60,
    86
  for women 2–5, 11, 29, 44–45,
    100, 121
  see also COVID-19
change, agents of 2, 9, 12, 15,
  77–79, 120–123
  fear of 76
  need for 3, 17, 24, 26, 29
  pace of 7, 47, 56, 62, 98
  structural 61, 64, 109, 115,
    117–118
children, childcare 44–46,
  104–105, 114
  child-free 12, 35, 51, 123
  desire for 27, 116, 122
  having 9, 13, 24, 40–45, 65, 99
  research with 18
  see also motherhood
chronic illness *see* disability
clusters, disability/chronic illness/
  neurodivergence 73, 117
  first generation 73–74
  parenting 11, 73

for support 11, 27, 78–79, 84, 88–89, 115
collaborative autoethnography *see* autoethnography
Colwell, Rita 2, 58–59, 63
communities 29, 65, 92, 114–115
see also safer space
connection, disconnection 48, 97, 107
  emotional 20, 29
  need for 25, 71, 95, 100
  through WISC 13, 85, 89, 90–92, 101, 115
contracts 35, 38, 40, 60, 83, 88
coping strategies 97
see also wellbeing
COVID-19 challenges of 31, 44–45, 50, 52, 105–107, 119
  lived experiences of 11, 65–66, 84–85, 91–101, 104, 110–114
  timing 9, 19, 25, 79–80, 86–87, 122
creative writing 19–20
creativity 22, 85, 90, 92, 109
Crenshaw, Kimberlé *see* intersectionality

D

Daniell, Ellen 2, 58, 64, 115
disability, authors' experience of 9, 121
  charter 62
  cluster 11, 73, 88, 115, 117
  intersectional 6, 13, 49, 52
  participants 82
  in science 27, 56–57, 63, 67, 116–118, 122
diversity, lack of 3, 47, 60, 70, 91, 118
  need for 7–10, 26–27, 29, 53–57, 123
  research on 23–24, 25
  work on 52, 74–75, 91, 103, 114, 116
Draper, Emily 73

E

Embodied Inquiry 18, 120
embodied, embodiment 16, 21, 34

approaches 9, 22, 72, 120
perspective 16–17, 24
practice 38–39
stories 8, 13, 20, 90
emotional labour 44–45, 76
Evans-Winters, Venus 22

F

femininity 12, 91, 101–104
feminism *see* intersectional feminism
fiction *see* creative writing
Franklin, Sarah *see* wench

G

gender
  in academia 13–14, 63, 86, 115
  bias 4–7, 63, 91, 118, 120
  boundaries 52
  caring responsibilities 11, 73, 83, 113
  challenges 26
  equality 7–8, 45, 61–62, 75, 120–122
  impact of COVID-19 8
  lack of balance 3, 6, 62, 70, 114, 122
  leaving academia 47–51
  marginalisation 4, 17, 26, 29, 33, 56
  in science 17, 27, 47, 56–57, 59–60
  see also bias, pressure
Goodwin, A. Lin 119
Gornick, Vivian 2, 17, 42, 58
graduate students, data on 3, 41, 47–48, 52–53, 55, 119
  support for 25, 37, 50, 67, 74, 95

H

Haynes, Cally 71
higher education *see* academia
Hiscock, Jennifer 28, 71–72
home, homeschooling 45, 65, 105
  responsibilities 42–45, 91, 111
  travelling to 6

INDEX

working from 45, 100, 107, 112
see also children
Hutchins, Kristin 73, 87

I

inclusivity
  drivers for 61–62
  inclusive approach 11, 21
  inclusive future 116, 122
  inclusive science 87, 88, 103, 117
interdisciplinary research 69, 78, 109
intersectional feminism 5, 29, 40, 52–53, 55, 119–120
  see also Black feminism
intersectionality 6, 29, 33, 61, 115
  see also intersectional feminism

K

Kieffer, Marion 71, 73

L

laboratory, accessible 73
  reality of 12, 20, 22, 85–86
  safety 91, 101–104, 111, 116–118
Lehn, Jean-Marie 69–70
Leigh, Jennifer 24, 38, 72–73, 85, 103, 121
LGBT+ 62, 63
long-distance relationships 88, 107, 122
Lorde, Audre 53, 77

M

Mahat, Marian 64, 115
marginalisation, building community 12, 92
  in higher education 13, 25, 38, 47, 52, 72
  marginalised groups 1, 9, 33, 75, 86, 87
  mentoring 28
  in science 6, 9, 17, 27, 47, 56, 60
  support for 2, 19, 61–62, 64, 74, 114
  women 3, 9, 60, 77–78, 115–116, 120–123
  see also gender, minority
Mason, Mary Ann 40, 42, 64
McConnell, Anna 71
mentoring, 63–64, 119, 123
  see also WISC
    mentoring programme
microaggressions 54, 56–57
Milano, Alyssa *see* #MeToo
minority, barriers for 40, 49, 53–55, 57, 60
  harassment 52
  supporting 60, 74, 114
  underrepresentation 12–13, 70
  vocal 26
  see also marginalisation
misogyny 5, 120
motherhood 3, 12, 17, 39–46, 58, 116
mothers 5, 33, 39–46, 105

N

networks, forming 37, 60, 67, 88–89, 114
  networking 52, 71, 101
  old boys 40, 88
  for support 37, 50, 64, 115
  see also WISC
neurodivergence *see* disability

O

obstacles 62
online learning 65
overwork 5, 14, 15, 33–38, 96

P

peer, group 28, 71, 83, 91
  mentoring 64, 71
  review 37, 59–60, 86
POWRE 61
pressure, early career 33, 51
  of academia 59, 85, 122
  from COVID 52, 65, 95–97, 111
  in a lab 117
  of parenting 45
  of running a research group 9
  of success 31
  for women 59–60, 107
Prince, Cali 105–106

publications, as measure of success 27, 31, 35, 58, 74, 123
in chemistry 26, 63, 70, 71
see also WISC

Q

qualitative research 10, 12–14, 16–20, 23–25, 87, 109
creative research 9, 18–19, 24–25, 72, 77, 86–87
on community 17, 25, 28, 71–73, 77, 81–84
on COVID-19, 44, 87, 107, 110–114, 122
surveys 10–11, 19–20, 45, 72, 91

R

racism 5, 26–27, 32, 53–55, 75, 116
rape 4–6
research groups, developing 37, 87
managing through COVID-19 44, 65, 84, 92, 110–113
reflective work with 19, 85–86, 91, 110
rhythm 77, 92, 119–120
see also rhythmanalysis
rhythmanalysis 24
Rosser, Sue 2, 12, 58, 61, 63, 64
phase model xvii–xx

S

safer space 25, 28, 64, 73, 76
seniority 35, 40, 84, 88–89
career stage 9, 25, 52, 76, 81, 113
lack of senior women 42, 49, 116
privilege 26, 76
senior men 13, 47, 68, 75, 88
senior women 51, 60, 104, 119, 122–123
support from senior colleagues 12, 27, 71, 77, 84, 88–89
sexual harassment 68
see also rape
Slater, Anna 73

social science 19, 23, 27, 35, 37, 72, 109
see also qualitative research
supramolecular chemistry *see* Jean-Marie Lehn

T

trans 2, 6, 88, 116

U

undergraduate students, data on 3, 47, 52–53, 59–60
experiences of 24, 50, 58–59
support for 25

V

vignettes, inclusions 31–32, 50–51, 67–68, 88–89, 107–108, 122–123
theory 8, 19–20, 39–40, 85, 90–91, 104, 121

W

Watkins, Davita 73, 87
wellbeing 33, 37–39, 46, 50, 122
wench 77
whistleblowing 1, 8, 19, 75
WISC, achievements of 13–14, 27–29, 77–84, 86–87, 109–110, 119–121
approach 16–20, 25, 85–86
board 24, 27, 73, 120
community 12–13, 32, 77, 88–89, 115
founding of 9–14, 71–75
logo 72, 78–80
mentoring programme 25, 28, 72–73, 77–79, 81–85, 88–89
online 25
publications 86, 121
women 13, 40–41, 44–45, 57, 64
see also gender
work life balance 37, 65, 81, 121
see also wellbeing

Y

Yarrow, Emily 40, 64, 91, 115